Kindred Spirits

Kindred Spirits

ONE ANIMAL FAMILY

Anne Benvenuti

The University of Georgia Press

ATHENS

An earlier version of chapter 10 appeared previously as the
article "Kindred Spirits: The Greatest Story on Earth?" in
Narrative Culture 4, no. 2 (2018).

Published by the University of Georgia Press
Athens, Georgia 30602
www.ugapress.org
© 2021 by Anne Benvenuti
Set in 10.25/13.5 Kepler Std Regular by Kaelin Chappell Broaddus
Printed and bound by Sheridan Books, Inc.
The paper in this book meets the guidelines for permanence
and durability of the Committee on Production Guidelines for
Book Longevity of the Council on Library Resources.

Most University of Georgia Press titles are
available from popular e-book vendors.

Printed in the United States of America
21 22 23 24 25 P 5 4 3 2 1

Library of Congress Cataloging-in-Publication Data

Names: Benvenuti, Anne, author.
Title: Kindred spirits : one animal family / Anne Benvenuti.
Description: Athens : The University of Georgia Press, [2021] |
    Series: Animal voices : animal worlds | Includes bibliographical
    references and index.
Identifiers: LCCN 2020044538 | ISBN 9780820359571 (paperback) |
    ISBN 9780820359564 (ebook)
Subjects: LCSH: Cognition in animals. | Animal psychology. |
    Human-animal relationships.
Classification: LCC QL785 .B368 2021 | DDC 591.5/13   dc23
LC record available at https://lccn.loc.gov/2020044538

With deepest gratitude, I dedicate this work to my two good mothers,
Corinne Julie Benvenuti and Sherry, Lady Bright III,
whose love and guidance have sustained me.

To Elizabeth Louise Hill, my sine qua non.

To Isaac Mathole (1979–2019), who lived and
died in his love for the elephants.

To Jane Goodall, whose pioneering work (from 1960 to the present day)
among the chimpanzees at Gombe, and now among human audiences
throughout the world, changed forever our understanding of human-
animal relations, of competencies and behaviors that we share with other
animals, and of the vital importance of animal sanctuary programs.

To Jaak Panksepp (1942–2017), friend and mentor.

To Fortunato, Jiminy Cricket, and Bunny, whose deaths
I attended during the writing of this book.

To Lexi, who sits beside me still, and to JouJou, Alpha, Tobia, and
Brezza, who look into my eyes, cuddle me, and remind me every
day of the simple power of touch, of play, of the vulnerability that
is the price of living, of the buffer that being connected is.

And to Centaura, who opened my old heart in a
fresh way, and who blazes there yet.

CONTENTS

ACKNOWLEDGMENTS ix

CHAPTER 1   In Times of Drought   1

CHAPTER 2   Hard Lives, Soft Landings   20

CHAPTER 3   The Big Five and the Little Five Million   40

CHAPTER 4   It Starts with One   59

CHAPTER 5   Endangered   76

CHAPTER 6   Oxytocin Bay   92

CHAPTER 7   Why Cats, Florence?   111

CHAPTER 8   Love Dogs   134

CHAPTER 9   Animal Ambassadors   155

CHAPTER 10   The Greatest Story on Earth   175

CHAPTER 11   Might We Yet Swarm?   190

NOTES   211

BIBLIOGRAPHY   219

INDEX   229

ACKNOWLEDGMENTS

I honor the kindred spirits—humans and other animals—whom you will meet in these pages, those who so generously shared themselves in bringing this book to life. I will always treasure the times that we spent together and the work that we continue to share, connecting humans with other animals.

I am grateful to the University of Winchester for the institutional support of this work, and for the many opportunities to discuss the ideas articulated here with colleagues from a range of departments and fields of study.

Special thanks to my literary agent Kathleen Davis Niendorff, who believed in this book and who persisted, and to Bethany Snead, editor at the University of Georgia Press, for her enthusiastic support.

Thanks, too, to so many colleagues that to single out anyone is certainly to miss others. Nevertheless, I am particularly grateful to my friend and mentor Jaak Panksepp, whose life work on the affective lives of mammals provided me with a platform from which to launch my own ideas. And I am grateful to Marc Bekoff for his broad and tireless work in animal research and animal advocacy, for our exchanges of ideas, for his support of my work, and especially for his educating and reconnecting of humans within the greater family of animals.

Revisiting the methodology of ethnography and the power of narrative inspired the shape of this work, and Sabina Magliocco opened the door to this by encouraging me to participate in American Folklore Society meetings and, with Michaela Fenske, by inviting me to write for *Narrative Culture*.

Grazie mille to Francesca Radin for helping me to interview Italian bee-keepers and for assisting with translation of the interviews.

Finally, I express deep gratitude to Elizabeth Louise Hill, who supported this work in many ways and over many years, and to Diane Koditek, excellent adventure buddy, who appears in this book more than once because she has so often been out there in the action with me.

Over the four years in which I was preparing to write this book, I visited with many people and animals in many places, attempting to understand the needs of nonhuman animals and the motives and methods of humans who work to save them. The stories herein are my own snapshots of moments in time. Any errors of representation are mine alone.

Kindred Spirits

CHAPTER 1

In Times of Drought

The little sparrow was panting in the shade of my front porch as I stepped out into the crackling heat. She was so weak that she did not resist being carried to the water bowl that I brought for her. Then she drank water, and she splashed in water—and life again! Yes, that simple. Water, and life again. In the southern Sierra Nevada mountains of California, where I lived for many years, the summers are hot and dry in the extreme, challenging wild animals to find water as the August temperatures hover in the hundreds and the creeks slow to a trickle. The sparrow's accustomed water sources had dried up, and she was desperate.

Not long after my brief encounter with the little sparrow, I added a new water bowl for the animals and birds who were suffering from the drought conditions then plaguing California. Another factor had also motivated me. It had taken me years to acquire an automated irrigation system, but when I finally did, I found that I had traded one problem for another. Now I was no longer having to get up early to water everything by hand before the heat evaporated every lick of moisture. But the squirrels and raccoons had quickly discovered that they could access a ready supply of water by chewing through my irrigation lines. Once I understood the nature of this new problem, I did my own version of chewing on it for a few days before I came up with an effective solution that involved neither poison nor predators, not even plastic pseudopredators. I put a large bowl made of temperature-resistant material under one of the big pines, dropped a dripper into it, and secured it with a rock so that every day the water bowl would fill automatically. And, no kidding, there was often a line at the water bowl, squirrels poised head down on the trunks of trees, birds perched in the branches, a fox camouflaged among the rocks, a feral cat bold in approach but quick to dart away. The drip lines remained safe while the bowl became quite the interspecies social center.

There came to this water bowl a raven whom I learned to recognize as a repeat visitor when he began to bring dried bread each morning. He would arrive after the morning rush hour at the bowl, just about the time that the dogs and I returned in the car from our morning trail walk. He would drop his chunk of bread into the water and knead it with his claws, then eat it. I'd pull the car into the driveway, roll down the window, and say, "Hi, Jason." Then I'd watch his bread-and-water dance. I came to see that his daily bread was always a bagel. Though he did not call me Anne or any other name I recognized, he knew me. We became accustomed to one another, the raven and I, and even the raven and the terriers. Jason soon noticed the large lava rock next to the bowl, a rock I had hauled from some far-flung place and set under the big pine tree. He began to stuff bits of his bagel excess into the holes in the rock, returning late in the afternoon for a snack. I was tempted to put out a salt shaker and a sugar bowl to see what would happen next.

Soon, of course, he was hanging out in the tree when not on reconnaissance, and as he grew, he began to bring his friends around. Jason and his bachelor buddies provided me with great entertainment. One day I saw multiple ravens hopping from one foot to the other, even dashing and flapping their wings at the black cat, who then sat down on her haunches, sipping daintily—the well-mannered and invited guest who, once sated, sauntered off, stopping to look back over her shoulder just once. Then Jason and the boys flapped into the water bowl. And, like Gabi Mann, the little girl in Seattle who has catalogued a vast trove of treasures delivered to her by birds, I received occasional gifts from Jason and his friends—a button, a piece of twine.[1] I did not know then to think of these as gifts, so I didn't collect them, and, worse, didn't appreciate them properly, simply because I didn't know to expect gifts from my raven friends. Perhaps I was expected to help myself to the raven version of toad-in-the-hole that Jason prepared each day? Maybe next time I have a raven friend, he will bring me gifts and I will gratefully collect them and say thank you, and I will feel rich.

I had not intended to solicit raven friends but simply to secure my drip lines from marauders in a way that was not hostile to my animal neighbors. I also hadn't meant to open a raven sanctuary, but open a raven sanctuary I had unwittingly done. One day, about two years into our friendship, I noticed Jason beside the water bowl, his beak open, his feathers ruffled. He looked lackluster from a distance. I let him be, knowing I have my own lackluster days. But when I came out a couple of hours later, he was still there, his sleek black feathers askew. As I got closer, I saw that he was panting and that there was a drop of blood on the dirt. If he spent the night there in that condition, I knew that the coyotes who come down the mountain each evening looking for sustenance would have him for early supper, just after dusk. I couldn't bear the thought.

Jason let me pick him up and carry him. Inside the house, where it was cooler, Jason pecked at the bits of bread that I offered him, and he drank water from a dropper, and he nestled into the spot that he chose, the tuck between my chin and my neck. I had no idea what was wrong or what to do for Jason. Back then, I knew very little of raven anatomy and physiology, so I could not imagine the needs of a sick raven beyond the basic biology and psychology that he and I shared: temperature regulation, hydration, a safe place in which to rest, allowing the body to focus its resources on repair and healing.

I had often watched Jason take flight from a branch in his pine tree, moving his wings to gain elevation until he became that wide V shape that is the hieroglyph for a soaring bird, way up in the distant, thin sky. Like all ravens, he was able to perform the mechanical kind of flight that uses the powerful pumping of his own muscles and blood, but like eagles and hawks, he was also able to soar. I'd seen him catch a draft and ride it, sometimes tumbling upside down midflight, or drop spiraling down and then pulling up in an impossible lift. He looked to be playing in the most carefree way imaginable, soaring in the air, and then shouting, "Look Ma, I'm faaaaaaallllling! Not really." Yes, I imagined it would be great fun, and also a test of skill, thrilling and satisfying all at once. But can I really imagine what it is like to be a creature made for that performance?

An adult raven is about two feet long, from beak to the last nth of tail feather, and has a wingspan of about four feet. Ravens in California average 1.72 pounds in total weight.[2] Their $27\frac{1}{2}$ ounces in imaginary distribution along their longest axis, the wingspan, averages out to a bit more than a third of an ounce per inch of length. My ounces (whose total shall not distract the reader), distributed along my longest axis, heel of foot to top of head, average out to just under three pounds per inch of length. I am bound to the Earth by gravity, while the raven is unimaginably light on his wings. How can he possibly manage this feat of physiology, being nearly as light as air, and assembled for living high off the ground?

How can he be smart like me, with his little brain the size of a walnut? My primate brain comes in at somewhere between two and three pounds. I grew up with the common assumption that my big brain made me smarter, but I now know that scientific knowledge of his raven brain has challenged that notion. It turns out that he packs twice the number of neurons into his forebrain, the brain region associated with executive function, than I pack into mine.[3] His cerebral closet is better organized, so he can fit more in a smaller cranial space.

But still I ask: How can he be full of feeling like me, relishing in play, as do I, and also be someone for whom mastery of a challenge is enormously satisfying, as it is for me? How can he recognize these features in someone of another species, as he surely does recognize my vocal play when I imitate his utterances, then vary them with increases in the number of repetitions, to which he responds in kind but with a single variation for me to hear and imitate, then vary? We do go on and on this way, sometimes at length. It surprises no hearing person to know that ravens are vocal animals, but it surprises many to know that they belong to the order of Oscines, the songbirds, named for their exquisite level of control of the syrinx, the vocal apparatus for making a range of sounds. While few would describe the cawing of crows as songlike, anyone who has observed ravens knows that they have different forms of vocalization for different purposes, from the loud caws used to draw attention to food to the guttural utterances, the clacks and rumbles of private conversation between mated pairs.[4] I once came upon a raven alone in a field, walking around on his bipedal ground apparatus, looking—and sounding—to me distinctly like a person who is talking to himself. I decided to test my impression by responding to him in my best raven vocal likeness. It was perhaps the most comical moment I have shared with a wild animal. He registered startled surprise raven-wise, raising up and cocking his head, taking a few hops in different directions, looking this way and that for the source of the intrusion into his solitary reverie. He was talking to himself and obviously startled to get a response, even more so from one of another species.

How, indeed, can we share all this while living in such different bodies, and such different environments? Me of earth, he of air.

He is entirely black, feathers, beak, feet, thus absorbing heat to warm him in the high, thin air. His bones are hollow to allow for the development of air sacs that fill as he inhales, and that move air into his lungs when he inhales a second time in succession. Yes, he inhales into his bones! The oxygen moves through his system continuously, even as he finally exhales carbon dioxide from his lungs, where the oxygen/CO_2 exchange occurs in him, as it does in me, when we in like manner send out the same chemical soup required for the plants in their respirations, the raven and I together inside that greater respiration of Earth, going about her metabolism. And though his bones are hollow, they are not proportionally lighter than mine, but heavier and stronger so that they can hold up to the pressure of those wind currents that he navigates with such virtuosity. He can eat just about anything, and does: insects, old carcasses of roadkill, restaurant trash, picnic fare. I'm a generalist eater, too, though unable and unwilling to eat in his range. I would reject in disgust the wormy, rotting fruit that is an extra-juicy and nutritious meal for him. His dietary habits also make the world a cleaner place as he goes about his "sky burial" work, storing small pieces of meat in caches that are often found and eaten by other kinds of animals.[5] Even though I am a relatively clean-eating vegetarian, I leave a huge trail of waste for the likes of the raven to clean from landfills.

I need to sleep in a nest, protected from predators and dangers of exposure to a variety of weather conditions. As my brain changes gear entirely for restorative sleep, my body must be still and safe. I so admire and enjoy a good sleeping nest! Surely, I think, ravens (and all birds, for that matter) must sleep in their gorgeous nests, made of branches, with soft inner linings. But no, they do not sleep in their nests. Because they have the capacity to rest one brain hemisphere at a time, the ravens do not need the deep sleep and immobility that I require. And they have a special feature in the tendons of their feet such that, when relaxed, the claws encircle and grip, assuring that the bird on the wire or the bird on the branch does not fall off when sleeping. I can scarcely imagine what such sleep feels like, attached as I am to the comfort of my own padded mattress.

We humans have been a distinct species for about 150,000 years, and they, the ravens, for three million years. We live on the same planet, in the same desert or forest or on the same farm, with shared fundamentals of physiology and psychology—ingestion, digestion, circulation, reproduction, temperature control, the impulse to play and to communicate, and to mate, and to master skills—and yet we inhabit our shared spaces very differently, in very different bodies.

Most of these raven facts are things I learned after knowing Jason and being caught out ignorant about how to help him. Though I wanted to help him, I had not bothered to learn much about him before I needed the knowledge, and then it was a bit late. So I did what I know how to do, the things that generally translate to other animals. I had done this before for Emma, an injured Hawaiian cardinal, and for Fling, a fledgling woodpecker who had flung himself badly from his nest, and they had both recovered and returned to their wild lives. I made for Jason a bed of old rugs in the garage with access to food and water, and I checked on him after a couple of hours. He was a bit better, but not significantly. I brought him into the house again overnight because I have learned from experience that safety is a good thing for wild animals who are hurt or just plain tired, as it is for us, and that isolation is most often not so good for them, as it is not so good for us either.

The next day Jason was not even trying to eat, and there were still occasional drops of blood where he'd been. Of course, I was concerned about contagion and was washing up like crazy after us. I invited my friend Diane, who has both knowledge of wildlife and a dependably pragmatic approach to life generally, to offer her opinion. As Jason rested listlessly in the towel now on her lap, she, too, was startled by his lightness of being. We decided it was time to consult the experts. I called the wildlife sanctuary fifty miles away and asked them what to do. Not surprisingly, they said that they couldn't diagnose over the phone and that I should bring Jason in. So I made him a deep towel nest, sufficient to contain his wings if need be, and got into the passenger seat of Diane's car. She drove the three of us down the long, winding canyon of the Kern River to the California Living Museum, where native fauna are rescued and rehabilitated or, if rehabilitation isn't possible, kept and cared for as they become ambassador-educators for the public who come to see them.

We pulled off the highway onto the dusty two-lane road that snaked through the parched hills until we reached the sign for the parking lot. I was grateful to be the passenger of someone who knew the way. Bypassing the entrance-fee gate, we went directly into the low reception building past the stuffed toy animals in the gift shop. I told the young woman at the counter that I had called ahead and had brought Jason. "Jason?" she asked. I laughed and told her that he was a resident of my front yard and I'd named him a long time ago. "Oh, you know the bird!" she exclaimed, fairly glowing with approval. I handed him into her careful hands, and we watched as she placed him inside a heavy cardboard carrying box with a handle at the top and big air holes around the sides. To be honest, it nearly killed me to see Jason go into that box. My friend Jason did not belong in a box, but in the sky, or in the big pine, or at his water bowl, cawing and cackling and clacking. I began to wish that I'd let him die at home, under his pine tree and next to his lava rock and water bowl. I was awakened from that brief reverie as the young woman told us he might have gotten into some poison, that the vet would come around and assess him that afternoon, that we should go home and call to check on him tomorrow.

Tomorrow, he was dead. Of course, I am not such an idiot that I don't know that animals die all kinds of deaths in the wild. Why had I not just let Jason die at home? The answer: Because he was my friend, because I hoped he might live to dunk more bagels on fine summer mornings, and, finally, because I know that coyotes eat *before* killing. Though I knew by the second day that he would likely die, I hadn't wanted Jason to go that way. I also know how deeply I don't want to die in a hospital. So I felt that anxious uncertainty that is common to people who want to help suffering animals as we do our best for them, often not knowing whether we have helped or even harmed our animal friends.

I did, in fact, see Jason again—or I think I did. When I was taking the wildlife rescue and rehabilitation course at the California Living Museum the next spring, a raven was produced from the freezer for us to examine, to get to know what a raven looked like and felt like, though frozen. "Jason!" I gasped, as I quickly explained how I'd brought Jason in for diagnosis. Of course, the staff rushed to tell me that there was no way to know that this one was that one. But seeing this dead bird, I remembered Jason living, what it was like to hold that bird who, though very large, was light as air . . . what his gently clacking beak sounded like so near to my ear, what it felt like on my skin. Yes, of course, Jason "beaked" me because a beak was what he had, but it was not fearful or aggressive, just a kind of handshake. I'd say you haven't lived till you've had a raven handshake, but it is not the kind of thing you can just procure. The raven has to want to shake your hand, and that can take some time. No online shopping will do it, and there will be no free delivery. You can't hurry love; you just have to wait, and you have to be a decent, respectful person while you are waiting. I've tried it both ways, and have learned there are no shortcuts to friendship with wild ones. They might need us at times, but they can't be bought. It turns out, though, that ravens generally do well in human environments, better, I might note, than some of the human inhabitants of the Tower of London, where the famous ravens of the Tower lawns have long flourished, living up to forty years there. There is an entire volume dedicated to the notion of the coevolution of humans and corvids.[6] If these authors are right about our parallel culture production, the next raven you meet may be more open to courtship than would be other kinds of wild animal.

After Jason was gone from the water bowl, other ravens continued to spend time there, and some came for help. Had they first come to the water bowl with Jason? Had the local raven culture changed on the basis of social learning? I only know that one raven needed a simple wing-wound cleanup, a safe place to rest, and some food and water. After two days, she flew off. Another had a paralyzed foot and seemed to want mere respite from the extra demands of living a bipedal life on one foot. He, too, left when he was ready to go.

Meanwhile, Jason had introduced me to the California Living Museum, and I returned there to become one of the volunteers who make the effort to know and love the local wildlife, and to create a bridge between them and the many humans who ignorantly—or innocently—cause them harm. And so the circle of knowledge and respect and affection grows, and so the sense of family includes more people and more animals who meet across all kinds of boundaries. I've come to experience it—and now to write about it—as a large-scale global cultural shift. Across the world, across cultures, a seemingly important aspect of emergent global culture is that people are feeling a new sense of connection to nonhuman animals as family, feeling their suffering in times of drought, feeling their delight at finding water, wanting to feel what it's like to be them, wanting to help them when they're in need. We're learning that we learn by listening to them, the other animals, that there are things that they know about this shared world of ours that we can learn from them, and that there are other things that we already know in common. And it's important that we feel it, because emotion is where we live. We want to soar, to swing from the web we have just spun, to hear with our feet, to see what we hear; we want to feel all the wonder of living, just as we'd prefer to avoid the pain of living. Emotion, after all, is not weakness, but an important way to evaluate experience and make intelligent choices.

On another dusty road, but this time halfway around the planet and three summers after Jason's death, I spot a bull elephant in the distance from my perch on the highest back seat of the safari jeep. At first he is just a gray blotch against the gray sky, moving toward the three giraffes we have been watching, then chasing them down, and finally lowering his tusks to the dirt and raising his rump in the air. In my excitement I call out the information as it happens, and Diane, who has come from California to meet me in Africa and is sitting next to me, simultaneously gives me a light version of "the look" while affirming my take on what is happening out there. Edward, the young ranger to whom we have been assigned, immediately puts the jeep in a low gear and edges up a gulley as he explains that I've just spotted a bull elephant display that means serious aggression may come next. I'm tickled to have spotted it first, deducing from his being alone that he is a bull, and reporting his behavior as it unfolds into what I now know is a musth display. At the same time, I understand from my friend's face that my exuberance about the distant bull may have frightened off wild animals closer to our jeep. Later, when I tell the story to Liezel Holmes, the head ranger, she simply tells me that my position on the high seat in the back of the jeep has a lot to do with my expert spotting abilities.

Edward had continued to drive away from the elephant before stopping to point out two young male lions who are feasting on a buffalo carcass, tearing meat from bone with great ripping sounds. These are lions of the lesser kind—that is, not the dominant male—feasting still on last night's kill. A buzzard sits high on the dead limb of a nearby tree, peering down, awaiting his certain turn. The lions don't mind that we are watching; the buzzard is patient. One lion moves from what was once the head of the buffalo around to the back, where the other is lying on the ground, systematically working the carcass; there is a growl, a snarl, as the one stumbles over the other. They feast together for a minute or two. Then the lion who moved first walks away, his stomach full, the taste of muscle and blood a lullaby for his long nap. Later, after dark, I spot two female lions, lazing heavy-lidded in the soft sand of a drainage. They, too, don't care that we have stopped to watch them; their eyelids flutter, one puts a huge front paw over her eye to block the light from the jeep's headlights. They also are in the languor of heavy digestive work as the night draws down.

The last time I saw a lioness, eight years earlier in this same place, she was crouched in the dry grass, invisible, golden fur within golden grass. Invisible, that is, until you see her eyes observing you, assessing. Then the little hairs on the back of your neck stand up to salute her. And the last time I saw the lead male of this group, he was just coming into his power, pushing his old dad out of the way. He was lean and muscular with his dark mane coming in, and he was making life hard for his father as the old man returned exhausted and dehydrated from his several days of work reinforcing the territorial boundaries of the pride. The current plethora of lions surprises me. Junior has clearly done well, if all of these young ones are his.

But on this second visit to the same part of the South African bush, I am even more surprised by what I don't see. I don't see giraffe, zebra, warthogs, antelope, kudu, all grazing together, the bush well-trampled and thick with animal life. I don't see elephants casually tossing small acacia trees without pausing to acknowledge the presence of visitors in a jeep. No, this time we go for miles without seeing any of the grazing animals. This is qualitatively distinct from how I remembered and wrote about it before.[7] I had seen then

many kinds of animals grazing together, not fearing—to my surprise—the lioness crouched down stealthily in the tall, gold grass. I tell Edward about my
memories, and I ask where the animals have gone. "Drought," he says.

Oh. That one terrifying word. I lived much of my life in California, where
the drought that began in 2012, sustained over several years, saw the collapse of underground aquifer systems in the San Joaquin Valley. The collapse was caused by human drilling, ever deeper, desperate and expensive,
to find water for the farms that drive the economic engine of the central valley. The drilling that took place during this drought caused the valley floor
to drop several inches every year.[8] But the story of the drought didn't stop
there. Tens of millions of California trees died on their feet, becoming fod-
der for the fires that travel fast on fierce winds.[9] It was a drought so severe
that not only did lawns and golf courses go without watering, but rivers and
streams turned dry and lake beds dusty. Water became a thirsty memory as
fires raged.

In the winter of 2017, there were massive flood conditions in California, spilling over the boundaries of streams and rivers and reservoirs. The
drought appeared to be over. But the trees were dead and no longer able to
anchor the soil, and the collapsed aquifer systems could not refill with water as they had done for millennia. Their walls and ceilings were no longer in
place. Mudslides closed major highways, sinkholes swallowed cars, and people representing many kinds of interests wondered if any of this water could
be held in the ground. This is weather writ large and contemporary: cycles of
drought and flood and fire met with short-term human technological fixes
that push beyond nature's capacity to restore and rejuvenate. In the summers of 2017 and 2018 an unpredictable and explosive kind of wildfire, driven
by fire tornadoes, created the worst fire seasons in ten years. In both years,
despite the official ending of the drought, people and animals alike lost their
lives to hot, gusting infernos.

In Africa as in California, Liezel tells me, the animal populations are de-
pleted; here because it was a drought year in which the normally cool winter never came, air temperatures remained high, and there was no water to
drink. The watchful ranger's eyes go a little dull as she says the words that indicate that this is abnormal weather, climate change on a grand scale. Yes,
she says deliberately, she thinks this drought is a new kind of thing, not just
a normal variant. The buffalo look visibly emaciated compared to my own
memory of them as fat, belligerent, sleek, and very many. Now they are few
and far between, seemingly dull of both coat and wits, and scattered, three
together at the most. They are the main reason that the lions are so happy

and prolific, it turns out. At one soft-sand drainage we see the dominant male lion with four females and five large juveniles napping languidly in the hot morning sun. After a few minutes of our sitting in the jeep nearby and watching the young lions, they begin to yawn, showing great shining fangs, and they stretch their giant padded front paws out in front of them, leaning into the stretch like golden-furred yogis. Then they saunter off, satiated and rested, not a care in the world.

About the glorious condition of the lions, I now understand that it is in fact related to the dearth of grazing animals. First, drought weakens the grazers. Second, when the big predators are eco-dominant, the herds tend to scatter instead of engaging in their protective clumping behavior. Where there was safety in numbers, there is now fear, and it's every grazer for himself, a desperate tactic that does not work for those whose protection is in the herd, in the belonging. Then the grazers, even the big and fearsome buffalo, are easy pickings for a thriving pack of lions, and the scattered animals live in deprivation and anxiety until such time as the waters return and the herds replenish. Or until the lions have nothing left to eat. When that happens, it can only go one way; then the lions will compete with each other in their desperation for food. Such is the cycle of scarcity in which death spreads like fire until all the fuel is gone and there is just dust and ash in its wake.

The next day, again in the high back seat but this time in another ranger's jeep, I point to the long furrows along the length of the roadside and to the tubes next to them, asking if the reserve is putting in irrigation for the thirsty animals. Just as I am poised to ask if supplying water doesn't cross the line from conservation to rescue, Aneen, our new ranger, tells me I've misunderstood the scene. The elephants, she says, have torn up the water lines that supply the human camps because their watering holes are dry, and they feel the water under their feet. I recall for a second the raccoons and squirrels chewing into my own drip lines in the California mountains. Aneen volunteers that elephant feet are very sensitive and protected by thick pads. I know that the elephants are actually hearing the water with their feet: quite literally, that they listen with their feet and so know where to tusk for the lines they pull up from the ground with their strong and agile trunks. I offer the information that they have something like sinus cavities in their feet, which are neurally connected to their ears. "Yaah," replies Aneen in her soft Afrikaans accent, "they do."

It would be convenient to think that the pendulum will swing back, the grazing animals rebound, the lions work harder, and evolutionary forces hone everyone's skills. Then the herds would become thicker and quicker, the lions might become more powerful and stealthy, and so on. But this is likely not the case, any more than it is likely that the collapsed aquifer ceilings in California will lift themselves from where they have fallen to allow for the storing once again of water sufficient to irrigate the once-rich valley soils.

The *Living Planet Report* summarizes decades' worth of scientific observation and analysis with its finding that 58 percent of the world's wild animals have been lost (and not replenished) between 1970 and 2012.[10] The expectation extrapolated from these several decades of data gathering and analysis is that 67 percent of the 1970 totals of all wild animals may have disappeared by 2020.[11] This half-century decline is not business as usual in the animal life of the Earth, but a description of the sixth extinction, now well underway and undermining the top predators of the animal kingdom who depend on the long-established patterns of plant and animal interaction, and on the interaction of these with the seasonal wind, water, and soil conditions.[12]

"Top of the food chain," I heard so many of my former students exclaim about us humans over recent decades of massive ecological change. In my own heart and mind, I belong to a different food metaphor, a table at which everyone is sacrifice and everyone feasts, and all are a small part of something bigger than the cycle of individual living and dying. All belong to life itself. In their unwavering pseudoscientific pride, the students never seemed to take in the indisputable fact that the top of the food chain is entirely dependent on the bottom and middle of the food chain. While it is psychologically appealing to identify with the powerful and few at the top of the food chain, it also risks losing touch with the reality that all the positions in that chain are necessary to life ongoing. And a very important aspect of this shared reality is that there is much greater mass at the bottom of the food chain than at the top. It takes a lot of bugs to feed the birds who feed the cats . . . and so on. Take away the bugs and the whole thing collapses.

The forces of deep interdependence driving the sixth extinction do not know that climate change is dismissed as a hoax by those who want to keep using the "old" ways of exploiting resources, like drilling for oil and mining for coal, like growing monoculture food crops such as corn and soy, crops and practices that destroy plant variability and adaptability. These so-called conventional practices are actually sudden innovations in the life systems of Earth, and they are destructive of the systems that are genuinely old and conventional, like rain filtered through rock and held underground. And the geophysical systems of soil and water do not know that human economies are apparently more important than the ecologies on which they depend.

Even if they did know, and wanted to cooperate with human economies, they do not have the capacity to balance the nitrogen from fertilizer that is dumped into streams and rivers and oceans, fertilizer used to grow food for eight billion of us, but also to keep (a few of) us entertained on bright green golf courses. As a consequence of these kinds of human ingenuity, the top of the food chain has grown heavier than the bottom, a model that simply can't work for long and must of necessity collapse.

It brings to mind Martin Niemöller, military man turned pastor, noting how the Nazi pecking order systematically removed people from weakest to strongest. Perhaps this is how it is now: first they came for the insect pests, then for the amphibians; now they are taking the fish (and the oceans) and the mammals (and the forests); and when they come for us, there will be none left to save us, nor to show us the way. The leading causes of this new mass extinction are habitat loss and degradation of natural systems, especially by way of climate change, building and development, and chemical pollution. My focus in this book is on the animals and the humans who interact with them, but I must put both in the context of Earth as a whole because we really are one whole composed of interdependent systems. To value any one part of this whole to the exclusion of other parts is folly.

But as a kind of counterpoint to our rather dismal context, I will be talking about the other animals, and I will be talking about the increasing numbers of people all over the planet who care about them and care for them. I rejoice, frankly, in the source of joy that humans and nonhuman animals are becoming for each other, in a variety of settings, as we together discover and extend our sense of being members of an extended family. I hesitate to repeat the facts of the sixth extinction and its causes because they so easily overwhelm us. But to ignore these foundational facts would be for me to add to the lies, and I would be whistling in a wind that would soon drown out my voice. Besides, what I am seeing in human relationship to other animals may be all the more revealing and remarkable for the fact that it is taking place in such a seemingly hopeless context. It is almost as though the great herd of animal life draws together in some instances, affirming our familial connection; and, in this new kind of herd, we give and receive comfort and perhaps ultimately protection instead of dispersing and scattering to our various fates.

It is my intention in this book, then, not to ignore but to limit the extent to which I speak directly of the massive scale of the abuse of animals, especially in mechanized farming industries but also in scientific laboratories, in poaching and the exotic pet trade, and in entertainment. There is already

plenty of bad news to be consumed, and it is not difficult to find. I am not interested in reporting ever more examples of how to be brutal, and more brutal yet, in the horrific hope that one might become hardened enough and
deadened enough to breathe the dust of an emaciated landscape in which
everything else is dead. No, it is the humans who draw together in solidarity
with other animals—and the animals themselves—who interest me. These
are my kindred spirits, human and nonhuman. The suffering will of necessity
come up at times, as we meet the people who rescue and give sanctuary and
also as we meet by name some of the animals who have been rescued from
abuse and rehabilitated. But my course is one that I trust will engender hope
and ignite the warming heartfire of care and comfort: a clear and easy decision, from my perspective, because there really is good news to inspire, educate, and motivate us. Indeed, meeting these animals and people might help
us to hang together herd-wise, enjoying one another and learning from one
another. Together, their stories make for many a fine oasis in the landscape of
seeming dust and ash. And who knows what might grow in such oases?

What's new is that there are people across every human category of political and economic condition, people of all religious practices and races and
genders and nationalities, and from many walks of life, who love nonhuman
animals passionately, who work to rescue and restore them from abuse and
from suffering. We are creating together a new form of global culture, one in
which the sense of belonging to the animal family grows stronger. There are
also those with entrenched economic interests in mechanized farming or in
raising animals for other kinds of profit, including lucrative trophy hunting,
who want to block our awareness. There are those who lobby for legislation
making it illegal for others to film their business practices, because of what
happens when we see where our meat, or cosmetics, or fur, or exotic pet bird,
comes from: we no longer want it. When we see the pampered trophy hunter
in his—and sometimes her—vanity, we detest them. Witness Cecil, the lion
killed by an American dentist who had to close his practice and go into hiding when exposed. This, too, is humanity: that we rescue, that we bind the
wounds and heal the terrors, that we don't want the meat that is the product
of cruelty and we don't want the neighbor who glories in bloodshed.

There are humans who work to preserve habitat and a natural way of living for wild animals, and humans who try to rebuild and restore wild populations, and humans who take in the animal victims of human development—the almost roadkill and the nearly poached—and who then attempt
to restore them to the wild. There are humans who rescue dogs from prof-
iteering dogfighting gangs and puppy mills, humans who feed and warm

(and neuter!) feral cats, humans who work to rescue animal companions when disasters strike, even humans who rescue the already rescued when they are endangered. Take, for example, the prisoners in Lancaster, California, who rehabilitate troubled dogs so that they might become adoptable friends, and who quickly took in forty-seven deaf dogs from a nearby sanctuary when it was under imminent threat from a raging wildfire.[13]

My long observation is that we humans are anything but simple. We are not heroes on one side of the fence and villains on the other. No, we're complicated, and we choose which kind of human we will be from places deep within us, but importantly, too, from the menu on our particular social landscape, from what we see and hear as possible. Within the constraints of each human life is a margin for choosing who and what we will be in response to the world around us. I try to choose to know people who make me proud to be human, who show me how it's done and who inspire me to do it, who brand my heart with sparks of courage. I return to places I have longed to see again and meet a new generation of animals as persons, each with a story. And I hope to do my small part to create a social menu with some good options on it, a menu befitting my shared table-food metaphor.

One more thing: For the past several years I have argued against what I call the old meme, the one I grew up with and the one that still reigns in many places—the twinned notion that humans are distinct from all other animals and also superior to them.[14] That two-part notion is wrong. It is scientifically incorrect to suppose that humans are distinct from all other animals in any way that would not also be true of every species in relation to all others. While it is certainly true that speciation is an indicator of unique adaptations and niches, this is true for every species. So, yes, dogs are unique among animals, as are elephants, and pangolins, and humans. But all are related and all share a great deal of similarity depending on peculiarities of adaptation and niche. Even more surprising is the fact that the species that are similar on a given trait may not be the ones we first expect to see. Instead, as we learn to pay attention, we may see another kind of animal who at first sight looks to have nothing much in common with us, but in whom we suddenly recognize important similarities. Humans, for example, are far away genetically from birds, but we share more language ability with songbirds like the raven than we do with great apes who are genetically very close to us.[15]

No animal is superior to all the others, but each kind of animal is best at doing what her environmental niche requires. There was a time in the not-so-distant past when evolutionary biologists dismissed other animals because they held an incorrect belief (the biologists, that is, for one distinction between us and the other animals is that the other animals do not make false inferences) that other animals are empty inside, and this belief kept them from asking whole sets of questions, much less finding the answers to those questions.[16] And for those who would dismiss the scientific view in favor of a theological one, who think first of creation and creator, why would humans and not the others also be God's beloved children? Why, indeed, except that it suits us to single ourselves out for divine affection, as we make a matter of pride our place at the top of the food chain? Even those who cherish the Bible and believe that it singles out humans as the only godlike beings in the vast creation are invited to look again, to question whether that is really what their scriptures say.[17] To think that we are distinct from and superior to all other animals is not just an innocent theological and scientific inaccuracy. It is a morally misleading concept, used to justify perpetration of wrongs on other kinds of animals as though they were mere things, resources for our use, rather than agents in their own contexts.

Our fundamental mistake in thinking other animals are things for our use allows us to do wrong to those other animals, but it also allows us to barrel along undermining our own lives. It is this kind of wrongheaded thinking that allows us to create the many kinds of ecological collapse that pull the rug out from under our own feet. When someone tells me I'm wrong about this, that humans are obviously superior, pointing to skyscrapers, or to satellites or antibiotics, or to our comfortable houses and video games, our convenient packaged food, and then perhaps recites the "top of the food chain" mantra, I point to instances of ecological ruin, from medical superbugs to the collapse of honeybee populations, to nutritional and attentional deficits, and to the ability of a coronavirus to bring the everyday human life of entire nations, indeed of the world, to a sharp halt. Then I suggest that a bigger view, one in which we are not the only valuable beings, would give us much-needed perspective and actually leave us living and feeling better, too, even to such an extent that we might need fewer antidepressants and painkillers.

If we have inherited a set of ideas that is both clearly and conveniently wrong, we have to be willing to replace it with a set of ideas that is more correct and that facilitates the good in us. In place of the "distinct from and superior to" meme, I argue that a scientifically correct notion would be to understand all animals as members of an extended family, or as kin. Even a basic grasp of biological science shows this to be not a fanciful notion but factual reality. Thinking of ourselves as members of an extended animal family is also morally good: it engages our concern for the others instead of allowing us to hold to the largely unconscious and self-serving belief that they are objects for our use, that our species contains the only real subjects on the planet.

Scientific observation tells us that other animals are not mere resources but are agents with their own interests and capacities, in many ways similar to our own: they want to live and grow, to develop and use their skills, to relate to their kind, to exercise their curiosity about the world. In some cases—as with Pacific gray whales—it is clear that they are as curious and interested in us as we are in them. People in the fishing villages of the Baja lagoons, where these whales come to birth their babies every winter, say that the whales have come to teach the humans forgiveness. Having spent time with those whales, having had whale mothers show their beautiful babies to me, I would say it is clear that they, too, are sometimes edified by cross-species relationships with us. Saying to ourselves that we and they are extended family is good science, and it's also good poetry that edifies us, that lifts our eyes above the horizon of the daily grind, and that may even allow us to gaze on an alternative and better future together.

The revolution in human-animal relations is happening all over the world and across interest groups. In some small pockets of the world it is perhaps not a revolution at all to think that all animals are an extended family of life on Earth. To my knowledge, most indigenous people understand non-humans to be other beings who share our lineage. When I was visiting the Amazon River in 1996, I asked an indigenous man, via an interpreter, if the people of his tribe ate the pink river dolphins. A look of horror came over his face. The response relayed to me was, "We don't eat people." Yet I had seen in the same village a mother agouti grilled for dinner while her infant watched from his woven cage. Clearly, this Amazon tribe also had some system for organizing a hierarchy of beings. Perhaps we find it easier to justify assigning personhood to large and complex mammals who are more like ourselves, and also dogs and cats and horses, our companions, groups not among those we typically eat. Apes and elephants, dogs and cats, dolphins and whales? Save them! Cows? Chickens? Pigs? Not so much. We humans are complicated. "Other animals eat animals," people tell me all the time. It's true, but while there may be room for debate about meat eating per se, there are no grounds for arguing the ethics of factory-farmed animals.

Jane Goodall and Marc Bekoff comment on this paradox, in which many of us humans participate, when they write, "Thousands of people who say they 'love' animals sit down once or twice a day to enjoy the flesh of creatures who have been utterly deprived of everything that could make their lives worth living and who endured the awful suffering and the terror of the abattoirs—and the journey to get there—before finally leaving their miser-able world, only too often after a painful death."[18] Yes, all of that is true. No need to look away, and no benefit from looking away. This really is the only place from which we can go anywhere else. Instead of looking away, explain-ing away, and carrying the burdens of uncertainty, far better that we con-sider with clear minds and open hearts the animal world, both around us and within us.

Back to the African bush, and now in the company of our tracker, Fossie. He signals for Aneen to stop the jeep at the juncture of two dirt tracks, far from any settlement, and he stands over a large stone on the ground point-ing to it. Do we know what it is? His eyes light with that now-familiar twin-kle. We don't. Then he begins to move some of the dirt from around it, turns it upright, and reveals the shallow indentation in which humans who lived in the old ways ground their plants for cooking. As soon as he turns it upright, I know what it is, and I move to it, holding my hands as if they were hold-ing an implement something like a rolling pin, and moving it across the sur-face of the stone. He smiles and positions his hand as though it were holding a pestle that he moves in a circular motion, telling me that I'm both kinda right and kinda wrong. Then he bends his powerful bulk and lifts the heavy stone, saying that he doesn't want anyone to remove it for their garden-design project or archeology museum. He uses his wide hands to make a hole for the hollowed stone at the base of the tree, and invites us to help cover up the sides. Diane bends down to distribute debris she's collected around our newly tamped-down earth, and we are soon satisfied that the stone is disguised again.

When I first met Fossie upon our arrival at camp, he was holding a tray in one hand, tongs in the other. As he presented us warm washcloths, I had noted his reserve, his quality of being contained within himself and not avail-able for my examination. And then, over our few days together, we all shared a warming as we came to know one another while also becoming acquainted with the land and the plants and animals, traveling on dirt tracks together in the jeep. He opened to us his knowledge of the African bush, broad and deep, intimate and integrated, and we soaked it up enthusiastically, telling him in turn some of the parallels we knew in the American West.

This day of moving and disguising the grinding stone was to be our last day together. As we stood appreciating our work, Fossie explained that he had positioned the stone so that it will catch the rainwater as it runs off the tree, and then the little birds will come to bathe. Then Fossie, middle-aged tribal leader, keeper of the old ways and the old knowledge, the same Fossie the modern tracker who works from the front of a big jeep, speaking English for his day job and translating his knowledge into the terminology of modern biology, Fossie, the robust and muscular father who is sad that his son is more interested in social media than in knowing the land or the traditional ways . . . Fossie wriggles in pleasure for us, in perfect imitation of a small bird bathing. Clearly, Fossie understands something of what it is like to be a little bird, happy to have found a little water in a time of drought.

CHAPTER 2

Hard Lives, Soft Landings

My friend Bill is a self-identified art-school bad boy of long duration. In difficult moments, he longs to be in his rather unusual happy place, in the company of Skye, Benedict, Dotty, and Louise, his friends who live in the sheep barn at Farm Sanctuary. Bill's hard living seems to call for some soft landings and the softest landing he knows is in that sheep barn.

I spent a rainy Sunday afternoon in the sheep barn with Bill and with the sheep and goats, and this is what I learned: that it just pulls on your psyche, the safety, the clean straw smell, the gently ambling warm wool tumbleweeds, the way the animals nuzzle you and gaze into your eyes. The way they include you. No doubt, this place is cleaner, fresher, just nicer than the average sheep barn because it is a place where people visit to meet farm animals, to get to know them as someone, not something. Imagine how you would keep your house if there was a *House Beautiful* tour taking place every day! But for all its unusual qualities of cleanliness and order, peace and safety, Farm Sanctuary allows humans to have the very fundamental experience of being an animal among other animals. This is an experience many people find to be deeply satisfying, so surprisingly satisfying that it frequently makes them weep.

I met Bill in person for the first time in the context of my visit to Farm Sanctuary, a visit he facilitated from start to finish. On my connecting flight from Chicago to Detroit, I realized that I had told my family only that I was going to visit Farm Sanctuary in New York. I had traveled a lot that year and gotten lazy about sharing my itinerary, and it suddenly occurred to me that I was going to an airport and city where I'd never been, planning to be picked up by a man I'd never met (in a black BMW named Maleficent) who would then drive me to his house for the night, and on to Farm Sanctuary in the morning. "How impossibly stupid can you be?" I heard my brain declare. I suddenly thought that I had perhaps been taking my self-directed lessons in trust and expecting the best in people a bit too far. I decided that I would send messages to my family containing the tedious details of my travel from the Detroit airport. After carrying out this small but important task, I caught my flight.

At New York's Stewart International Airport, just outside the terminal, I saw the hard-to-miss Maleficent, a BMW 428I xDrive sports car that I would normally take in simply as a flashy car. Next I met the real Bill Kauffman, who immediately struck me as sophisticated, well groomed in that used-to-being-good-looking way, with an air of being simultaneously above it all and perpetually ruffled by it all. I knew that Bill was an art school graduate and retired professional photographer, so I did perhaps read into his seemingly cultivated self-presentation. I'd met him on social media and knew him to be an animal activist, a vegan, a photographer, someone who made a lot of jokes about drinking, and a man obsessed with Farm Sanctuary.

We did, in fact, go to the Victorian home he shared with his wife, Patt, a Lutheran minister who would be going to work the next day while Bill and I carried on to Farm Sanctuary. The three of us shared a long conversation at a local restaurant. And I got a good night's sleep before heading out to Watkins Glen the next morning with my chauffeur, Bill, in the classy comfort of Maleficent. On the way I introduced him to dill pickle potato chips as we drove through the gold and orange and red late October morning. It was a longish drive, inclusive of two dill pickle chip stops. We had plenty of time for getting-to-know-you conversation.

I knew that Bill was proactively a vocal vegan. I, at this time a vegetarian who still occasionally ate fish, sincerely felt like a moral inferior. I asked him how and when he'd become vegan, expecting him to recount the degrees of abstinence by which he arrived at completely vegan. He told me that in 2012 he'd clicked a link in an email his daughter had forwarded, opened the video *Farm to Fridge*, and soon found his senses inside an abattoir.[1] He told me that he'd screamed "NO!" in horror, and he slammed shut the computer and flew out of his chair, changed forever. He wanted nothing more to do with the human exploitation of animals, ever. The man who sat down that day was one who enjoyed steak, the man who stood up was vegan. That had been four years before our scenic drive. Once Bill had learned how fast and how radically his behavior could change, he decided to quit smoking too. His doctor congratulated him and suggested his health might improve with a bit of animal protein. Bill explained to his doctor that vegan wasn't a health choice for him; it was a moral choice.

I told Bill I was somewhere on the way, that I still ate fish once in a while, but expected those days would soon be gone. I knew of course that eating fish was an artificial choice to change my own behavior by degrees, since fish are no less sentient than other animals.[2] A fish loves her life no less than does a cow or a human. But I knew by way of my training as a clinical psychologist that behavior change rarely happens the way it did with Bill, that sudden conversions most often also take sudden flight. Most people make lasting changes a bit at a time, or after many attempts. I quit smoking four times, and the last one took. The fourth time I quit smoking was more than forty years ago and I've never wanted a cigarette since then. But I had the advantage of having practiced quitting three times prior to the one that actually took. That might sound lame, but baby steps and relapses are part of the process of behavior change for most people. My mother, however, who raised me on the virtue of willpower, is not one to not be able to quit anything. She had said for several years that she wanted to want to be a vegetarian, but she did not yet really want to be a vegetarian. On New Year's Day of her eighty-fifth year, she did want to be a vegetarian; the time had come, and she changed completely. She never made the fish distinction. She was just done eating flesh, after letting herself want to want to change without actually changing for a few years. There are different ways to take it by degrees, but by degrees is how most of us change, and part of that process seems to be figuring out what actually needs to change.

When I had decided the previous summer that I needed to visit Farm Sanctuary, I knew I would ask Bill, passionate advocate for the place, to help make it so. For his sixty-fifth birthday, Patt had taken him and his daughter and son-in-law, career animal activists, to Farm Sanctuary. Here he'd met Susie Coston, Farm Sanctuary's national shelter director at that time, and had received the "Susie tour." He had fallen in love with Farm Sanctuary, and surrendered to the dawning realization that he preferred most animals to most people. But not all. Susie Coston remains pretty highly placed on his list of favorite beings. At Farm Sanctuary, Bill experienced animals as persons, one chicken at a time, and one turkey, one pig, cow, goat, and sheep. It changed him again. He was immediately comforted by the animal persons he met, and comforted by the fact that there they were . . . persons. Friends, even, friends who remembered him, welcomed him, hugged him in their various ways. In their vulnerability, he could face his own.

By the time we'd checked in at our somewhat eerily untended hotel, it was late in the day, so we decided on a drink, dinner, and an early retire to read and sleep. We sat downstairs on period-style furniture upholstered in dark fabrics in a room that looked like it would have been called a drawing room, though it might better have been called the family photo gallery. I found it kind of creepy to sit in someone's living room surrounded by photos of the entire extended family but without any signs of life. We sat peering out through several layers of lace that hung inside the windows at the darkening gray sky and wondered if the weather prediction of sun the next day could possibly hold. I'd gone out in search of a bottle of red wine and, as I sat to pour, Bill reached into his camera bag and pulled out a Nikon and a bottle of Elijah Craig bourbon. Bill claimed that Elijah was an underrated gentleman, whose company he found soothing, but I found Elijah to be a bit of a strongman, even if pretty smooth. It snowed overnight.

The next day we had a short and sunny drive to the rolling hills, red barns, and white fences of Farm Sanctuary. Bill was beside himself with excitement, like a five-year-old on his way to Disneyland. When we arrived, Farm Sanctuary looked as I'd been led to expect, based on—of all things—a critique of its very perfection. I knew to expect a pristine version of an ideal farm because I had heard that it looked like an ideal version of a factory farm, when it should perhaps aspire to look a bit more like a place governed by a kind of interspecies homeowners association. The critique I'd read suggested that perhaps farm sanctuaries should strive to distinguish themselves from factory farms—an interesting idea, I'd thought when I read it. After all, why replicate the place you want to tear down? Still, this farm was beautiful, simply beautiful. And perhaps the answer to the question about the wisdom of replicating an animal farm is that a beautiful farm can be a very good place indeed for farm animals.

Farm Sanctuary in rural New York had become a destination for me the previous June at the University of Winchester in England, where I had attended a seminar in the animal welfare department. Sue Donaldson and Will Kymlicka, authors of *Zoopolis: A Political Theory of Animal Rights*, were discussing both the necessity for incorporation of nonhuman animals into our political systems and avenues to accomplish the inclusion of nonhuman animals.[3] The question of whether humans are ethically obligated to be legally and politically concerned with animal welfare is no longer considered to be morally debatable by most philosophers and ethicists. However, expressing animal rights in human legal systems poses some really complex challenges because legal systems are by definition human institutions, derived from and serving human needs.

I had just written a paper, about to go to press, acknowledging the duty to incorporate animals as other kinds of persons into legal systems and discussing the complexities of doing so.[4] A fundamental problem is that current laws protecting animals are based on the notion that nonhuman animals are property deserving of protection only as such, a problematic foundation for claims of any kind of legal sovereignty or inherent rights.[5] Further, democratic systems of law require the consent of the governed, something animals cannot give, as far as we know, and may well not desire to give, were they able.

Sue and Will had made the case for inclusion of animals, especially companion and domestic animals, such as farm animals, on the basis of their being stakeholders in local communities. They argued also that these animals do express preferences, and it is incumbent upon humans in the community to listen and to respect the expressed preferences of other kinds of animals when articulating systems of rights and obligations. I thought that their argument was clean and pragmatic, though it would take quite a bit of work to begin the process of including the preferences of nonhuman animals into our political and material structures. It is not yet feasible to envision the dog parks being adjacent to the cow play pastures and the pig water parks in every town, but that doesn't mean it never will be.

Over the course of that afternoon of intense discussion and (to my surprise) truly delicious vegan catering by the University of Winchester, Sue and Will referred to a paper they had just published, suggesting that farm sanctuaries should move out of the rescue/advocacy model and into one of shared community of interspecies relationship.[6] Maybe the other animals don't want to be ambassadors to humans, maybe they don't want to be anything to humans, they argued, and that possibility should matter. In rescuing animals, caring for them, and asking them to be ambassadors to humans, the agency is still all human. But human policies should include the agency of other animals, say Sue and Will, who clearly think it is time to push into the new frontier of animal inclusion in decision-making: "But the offer of safe refuge is not the same as the opportunity to create a new and shared interspecies society. A different vision of a farm sanctuary would see its animal residents less as refugees and ambassadors, and more as citizens and pioneers of new 'intentional communities' who are given the freedom to create a new social world."[7] I left that seminar in Winchester with two commitments: first, to immediately recall my paper so that I could include reference to their suggested model for animal participation in human legal systems; and, with thoughts of interspecies communities dancing in my head like the proverbial sugar plums, to visit the original Farm Sanctuary in New York.

Four months later, as Bill and I turned in to a gently rolling driveway, past the white board fences and the red barn, I was thrumming with the sense of exciting possibilities, thinking that this was the place where an entirely different kind of world might be born. We parked behind an industrial building and walked past slop buckets and hay bales into a cavernous metal structure and up a wide and heavily textured metal staircase. No muddy boots gonna slide around on the stairs here! At the top of the wide gray stairs, Tara, right arm and chief assistant to Susie Coston, greeted Bill enthusiastically. She offered coffee and directed us to take a seat in the large entry room, where every window was occupied by a desk, and off of which we saw several closed doors. It was a standard office setup but it still had the air of a farm about it, literally smelling like animals and hay, with farm tools and buckets inhabiting nooks and corners.

Susie strode in, wearing cargo pants, boots, and a field jacket. She introduced herself and excused herself within the space of a minute. As manager of five farm sanctuaries across the country, she was a busy woman who exuded an unusual kind of energy, even wild, yet also very intensely focused. Indeed, there seemed to be something of the laser in her presence. She is not the holder of *a* grand vision, she is the holder of several grand visions, some beautiful, some horrible, but they are just context for her because her determined focus is on doing the work that is in front of her. In the course of our conversation she returned repeatedly to the theme that life and humans in particular are immensely complicated—and that, in our complexity, we are the makers of big problems for many species in addition to ours. She returned again and again to the simple fact that these other animals are real beings with selves and stories, that each of them is someone, not something, innocent victims of our complex capacities.

When she called us into the conference room where she'd taken the important call, Susie announced that she had one hour for an interview and then would turn us over to Tara for a tour of the farm because she had to catch a flight. I was invited to hang out with the animals after the tour, with Bill, the experienced farm sanctuary fan, the guy who knew the ropes, as my chaperone. Susie flipped her long straight hair back from her face, took off her field jacket, and asked someone to bring her coffee.

She had that "well, bring it on" attitude, and I started our interview on the wrong foot by referencing Sue and Will's farm sanctuary paper. My free-spirited interpretation is that she thought their notions to be pie-in-the-sky-when-you-die. Interspecies community? She was happy to keep the various animals from killing each other. Human-animal Shangri-La? "We've tried the cross-species thing—and it's the funniest thing. *National Geographic* calls us monthly and asks, 'Do you have any stories like a chicken and a pig that get along great?' That is absolutely adorable when it happens, but you are setting people up to fail because now everyone wants *that*, because it is so cute." She believes we are trying to make things too idealistic. "We actu-ally work hard to make sure the animals are with who they want to be with. But we've had lots of bad things happen when you cross species; we just had a chicken with a broken leg because she was living with a goat."

I asked her if there wasn't some vision to be realized in that model, even though it isn't practical under current circumstances. "I do think the ani-mals should have absolutely every right that they can have, but I also think there has to be some place where they are safe, and not just allowed free-doms that will destroy them. I've been working with them for twenty-three years. It's a lot different than walking through for an hour. Do you just let them have lots of babies? Because we have one or two and pigs have twenty." She laughs at the vision of a rapidly expanding world of pigs but then points out that "in the wild, most of the baby pigs die quickly in competitive condi-tions." I knew this. In fact, a majority of all the animals born in nature die be-fore reaching reproductive maturity. The most fit ones live to reproduce, and so the story of life continues to unfold.

Susie went on to explain why she often couldn't keep even animals of the same species together. She had to keep male and female turkeys separated because the white-feathered turkeys had been selectively bred to make both a lot of breast meat and thin, crispy skin for human gustatory pleasure, with-out consideration for the fact that such unnatural skin cannot do the job of protecting the females from being shredded during mating. Oh, well. "They destroy the girls, yeah, we can't even have the boys and girls together. So again, saying we would love this community where they're all doing their natural thing and making choices? They don't have the ability to do their natural thing. It's gone."

I asked her, if not into some kind of intentional interspecies commu-nity, where did she see it all going, what did she hope the condition of farm animals would be in two hundred years? She offered that it's complicated. The ideal would be to return the animals to their natural environments, but there is no place on the planet where these animals can be put back into their natural environments because they are not natural animals. "You would never intentionally breed a broiler chicken or an industrial turkey, not from the animal's point of view." She pointed out that we have similar problems with dogs and cats: bulldogs who can't breathe, Chihuahuas that must have C-sections to give birth to huge litters, dogs and cats of all kinds neglected and allowed to breed neglected babies—or worse, to be abused, objects of the anger and depravity of humans. "Look at the Holsteins, these giant Holsteins, they have hip and leg problems, they are *designed* to get too big. As they get older, they can't even walk up the hill. And they consume a lot of grass, an unnatural amount of grass."

Susie would be thrilled to put a stop to the monstrous breeding that created animals whose genetic manipulation is deeply detrimental to their lives, designed by humans to fatten quickly and painfully for merciless slaughter and meat production. "There are still animals that are closer to their natural condition, the heritage breeds. They still have their complications, they still have their problems, but they are closer to nature." She noted how few of the world's billions of farm animals there were, who are able now to live out their lives receiving human care and in as much peace as their genetically modified lives allow. She suggested that Farm Sanctuary does create a kind of ideal moment, somewhere between the horrendous factory farm and competition in the natural world, a moment wherein humans get to encounter these animals. Of this experiment, she exclaims, "Oh my God! Look at you, look at you, and you!"—and then muses, "But we don't really know where it goes."

Within fifteen minutes there was a knock at the door, an urgent call from another farm sanctuary about diagnosis and treatment of a constipated chicken. Yes, it sounds like it's meant to be funny, but it wasn't a joke, and no one here laughed. Broiler chickens are bred to be ravenous so they will fatten quickly. The goal is to move from birth to slaughter in forty-two days. These birds eat anything and everything: feathers, coins, nuts and bolts and nails, whatever they can get their beaks on. Farm Sanctuary took in seven hundred broiler chickens at once during Hurricane Katrina, and it was quite the learning experience. One of the chickens, whose crop had to be surgically removed, had a baby snake caught in the debris. Surgical removal of problematic chicken crops is common at this farm, because of course they get sick as a consequence of their inbred eating habits.

"Bar none," Susie says, "the girls are the sweetest birds you'll ever meet. When they're little, they will sit in a huddle in your lap. They love to be together. They like to sleep with wings over each other. They're the sweetest birds. The breeders didn't take that away from them." Susie's description of the pleasure that chickens have in sleeping together is echoed by Alice Walker in her book *The Chicken Chronicles*: "What do chickens like to do?" she asks in a chapter by that title. "They like to take naps! I would not believe it if I hadn't seen it with my own eyes. Not only do they like to take naps, they like to take naps together." Walker goes on to describe the chickens burrowing into the earth with a circle of their special friends and nodding off into "the most sumptuous quiet."[8]

But the broiler chickens on large farms are expendable by design. Susie tells me that about twenty million broiler chickens a year die of heart attacks because they are so fat. They never make it to slaughter, but no matter, because they are tax-deductible business losses. And they are only a fraction of a percent of the 8.2 billion chickens who do make it to slaughter each year, also essentially having never lived. As genetic engineering creates its inescapable problems, the factory farms create nightmare mechanical conditions that are escapable only for the very lucky few who are rescued. Even the farmers who operate large factory chicken farms sometimes find the work to be an unbearable nightmare. While visiting a factory farm and interviewing the farmer, neurologist Aysha Akhtar reports being told that at eighteen months the egg-laying hens begin to slow down and are then euthanized by being dropped into a dump truck to be killed with carbon dioxide gas. The farmer describes it this way: "They just start panicking and climbing on top of each other. Ninety percent of them die from smothering themselves before the CO_2 is even turned on." Akhtar was surprised by the way the farmer shared her horror. She asks him, "If it bothers you so much, how are you able to kill them?"

"We just shut our eyes and we just don't think. We can't bear to think."[9]

Some of the lucky few who make it to Farm Sanctuary are animals who dared to fight for their own lives. In dark and crammed gestation pens where sows are artificially inseminated and in the farrowing pens where they give birth and nurse their piglets, some pigs resist. In the midst of being packed into transport trucks headed for the slaughterhouse or at the very door of the abattoir, some cows resist.

Julia Pig was one such sow whose heroic resistance brought down on her both horrendous pain and amazing victory. Julia was a sweet-natured pig, but she stubbornly refused to move from the gestation pen to the farrowing pen. But the farm workers were required to move her to the farrowing pen, and when she wouldn't cooperate, things got increasingly tense. One of those workers captured on his phone what others were doing to try to move stubborn Julia. He used this audio recording to make a report to the New York state troopers, who came and removed Julia from that crate for entirely human reasons. They had heard the recording of Julia screaming and they saw the wounds she bore from the electric prods used brutally all along her body. After a particularly intense scream, one can hear a worker saying, "Oh yeah, I bet that one hurt!" Then two workers drag Julia by her ears in the direction of the farrowing pen. The police interfered because that kind of human behavior is illegal, and that kind of abuse of animals is illegal because we do care about the experience of other beings and we want to spare them suffering. As Susie so frequently points out, humans are enormously complicated.

The cops rescued Julia Pig, who soon found her way to Farm Sanctuary, where she quickly learned to answer to her own name. In spite of how she'd been treated by some humans, Julia Pig never showed any aggression to them, perhaps because of the kindness she now received from others of their kind. Maybe she was smart enough to somehow grasp the complexity of us. Or maybe it's that she landed on the right side of us. We just don't know much about pig minds because we haven't really asked before now.

Eight hours after her arrival at Farm Sanctuary, Julia started to give birth, and she kept giving birth through that night, to seventeen (count 'em!) piglets. Julia would spend the rest of her life with those seventeen babies, wallowing and pasturing in open fields and sleeping in deep, clean straw. After her death, Farm Sanctuary put together a video, *Julia: An Incredible Pig*, in which Susie Coston recounts the story of Julia. Susie begins by saying that the video was late in the making because there was a lot of grief to get through before anyone could face making it. Clearly, Julia had deeply inhabited a few hearts in her forever home. "The most beautiful thing about sanctuaries is we have all these families," Susie tells us through sniffles and tears, "and families that get to stay together. We really put so much energy into making sure that everybody was where they needed to be to be happy, and she was happy." Julia was a superstar pig, but one whose past caught up with her. The brutality had caused injuries from which Julia eventually died after several years of a happy life with her babies in sanctuary.

The eventual death of many animals from the very conditions that brought them here is part of the work. Many of these animals die from the effects of the brutality and deprivation they have experienced. Farm Sanctuary tries to let them be individuals—who have friends of their own making, and who sometimes even live in natural family groups, who have names and faces and preferences—so that they can be happy before they die, whenever that death comes. Julia left a legacy in her seventeen children. Susie concludes the video with these words: "The difference between Julia and all the other pigs in that facility is that Julia got out. Julia is special because we got to be with her, and the biggest difference is that we got to know her."[10]

My stomach starts to turn at the massive complicity that is my own ignorance in all of this. Because it is our ignorance, our capacity to look the other way, to just not think, that allows for the misery of billions of animals who think and feel, who have desires for their lives as we do for ours. We have come a long way, indeed, from the heritage breed cows giving milk for a family or even a village. And, since we weren't looking along the way when somewhere we made a wrong turn into the factory farming that creates billions of lives of extreme suffering, we have no idea how to get back to where we made that wrong turn. I think it is important to accept both our innocence and our culpability because it really, really is complicated. That complexity, I think, is not a reason to look away.

The big picture is overwhelming, perhaps even more so here than it is for wild animals. Susie recognizes several big pictures; she does not turn away from them. Yes, there are more and more vegetarians, isn't that wonderful? But she points out that there are also more and more humans, and more and more markets for meat. And we are so far from the difference that her efforts might make to the whole hidden culture of factory farming that she really can't anticipate the possible outcomes. Who knows? And so she intentionally keeps focused on the small picture, the one animal that is in front of her at each moment.

Taking care of the very few of these chickens selected by fate to become ambassadors to humans takes specialized knowledge and skill, and Susie Coston is a national repository of both. I made a mental note that this brand of contemporary chicken, bred for meat, is far away, indeed, behaviorally from the wild chickens I'd seen on the Amazon River in Peru, chickens who hang out near to human habitations, and who occasionally do become dinner for humans, but who are colorful and wily birds, full of themselves, the hens strutting and clucking along the forest floor with trails of chicks following behind them, roosters crowing from thatched roofs. Among those birds, the hens who didn't want to mate joined forces to thwart overly forward roosters.

It began to dawn on me just how far this show had carried on down the road since I was a child. My neighbors had kept chickens for eggs and for meat. I'd watched once in horrid fascination as the freshly decapitated chicken ran in a circle and the teenage boy who'd done the slaughter explained to me that the chicken is dead, but his brain was still telling his feet to run. The message was already in the nerves when the chicken died, my neighbor explained. And I recalled, too, the friend who grew up on a big Midwestern farm, artificially inseminating the cows and raising animals for Future Farmers of America competitions. She especially mourned her prize calf, Butch, whom she'd loved and cared for from birth. She spoke of the pride she felt when he won first place, and of the gloom she felt when she woke up to the fact that it all meant he was eminently sellable and would soon be gone from her care. She had also accepted payment of twenty-five cents a head to bounty hunt gophers on the farm. In her maturity, she deeply regretted these things that she had done as an innocent child. During the time I knew her, she never ate meat.

As I considered the ways in which the lives of farm animals had changed in my lifetime, I thought of the movie *Cloud Atlas* in which a waitress refuses to accept abuse from a customer and is removed from service as a consequence. We next see her enter a room in which former waitresses hang from ceiling hooks, ready to be processed as the meat to be served, all in a very mechanical, heartless, and eminently functional system. We see in the face of our waitress the dawning understanding as she recognizes former colleagues, and we see her decision to fight. "I will not tolerate this criminal disrespect" is her line to repeat from one incarnation to the next, as the movie flies through time. It's the kind of abstraction I can't imagine a chicken grasping in order to participate in the systems of justice that we humans have. I wonder if that's a limit of chicken minds, or a limit of our systems of justice, or both?

Clearly Susie Coston was not considering attempting to set up a parliamentary process here, or at any farm sanctuary. I feel deep sympathy with her entirely pragmatic and clear-eyed approach. Yet I still admire the work of Sue Donaldson and Will Kymlicka because the first step in any human undertaking is inner vision; we must have an image to lead us forward, and these ideas and images are creative works that, if they did fit with current realities, would not have the power to lead us into a different future. Yes, we've made the lives of farm animals complicated, and cruelly so, and what we do next is likely to be equally complicated. But I hope it will be less cruel. So here's to Susie Coston and to Farm Sanctuary, saving the few and saving them well, and allowing for a time and a place where humans can come to experience these animals, one by one, as someone, not something. And here's to Sue and Will, creating visions of a possible interspecies future.

Should the humans and the other animals have a future, to have a shared sensibility and structures for surviving together would be a beautiful thing. And really, why is any hope more wild than any other?

After Susie's intense and unfiltered intellectual clarity and emotional honesty, the farm tour was a welcome relief. We emerged from the metal office building into a sunny day with autumn leaves bright in warm colors dotting the hillsides and lining the lanes, with white clouds in a blue sky, with newly fallen and even more newly melted snow on the ground. First stop, the sheep barn, outside.

Someone rang a hand-held bell powerfully, and then sheep and goats began to flow down from an uphill pasture, past us and through the double doors of the beautiful old wooden sheep barn. This is where I first met Skye, a very tall goat, cavorting at the front of the parade, the leader of the goats. I never feel a hundred percent sure when someone points out for me a smiling animal, but Skye just had the most definitely twinkling eyes! We wedged through the door past them into the warm dark of the place, with its fresh straw smell, our eyes adjusting as big chinks of bright light spilled through the cracks between the old wooden boards of the wall. We were introduced to Louise, the head sheep, and then, up close, to Skye.

Here I also met Benedict Goat, as one of the volunteers was making ready to strap on his pink and black wheels. As we approached, Benedict dragged his lame rear legs out of the deep straw, and I watched as one of the farm workers attached his wheelchair to his body, cinching the nylon straps that give him two wheels to do the work of rear legs. Benedict's legs, by the way, just were that way, paralyzed by a genetic defect: it wasn't a particular act of harm that anyone did to him.

We followed Benedict out of the barn into an enclosed yard where white turkeys were walking and talking. You couldn't help but have your heart leap to see Benedict confidently step into his harness and then run, strewing a trail of joy behind him. He dashed out the big double doors into the grassy field, passing slower goats and sheep and turkeys, too, and then he tumbled over sideways when the contour of the hill and the wheels of his chariot didn't quite work with his speed and angle of descent. It was hard to see Benedict's joy turned to helplessness and failure like that, to see his front legs running in the air for a few seconds, as one wheel stuck in the grass and the other wheel spun free. But if Benedict's fall caused me to reconsider his wheels, the entire scene caused me to reconsider a great deal about living. Because, of course, someone came to his rescue and righted his wheels, and someone stood with him as he shook off the dust and dismay, and someone walked with him while his enthusiasm rekindled. From the depths of my own childhood moral training in Catholic schools welled up the thought, "Blessed, Benedict, are the meek, for they shall inherit the Earth." And blessed are those who gave you wheels, those who hunger and thirst after justice, for they shall be filled. And blessed, too, are the merciful, who witnessed your failure and put it to right without shaming you, for they will not be shamed but will be shown mercy.

Farm Sanctuary is one of so many places in this world where the poor hunger and are fed, where those who mourn are comforted—and, trust me, mourning humans flock here to be among the animals who know mourning. So I'm not just talking about the nonhuman animals, but also the humans who are here because they're poor, because they're hungry, because they long for justice and mercy, because they need comfort. Bill had told me that he cries, no, that he weeps, every time he comes here. And I too have turned to nonhuman animals for comfort for many years. But my afternoon in the sheep barn was something else again, something that led me to understand the comfort of a farm sanctuary more deeply. Susie's work is for the here and now, and maybe it will do something for the future also.

Tara and Bill and I walked toward the pig barns just as the pigs were being called to a late lunch. These animals were huge! I had not seen a farm pig, other than a heritage breed pig named Onion on an organic farm in Wales, in thirty years. These pigs were three or four times the size of that heritage pig. Because it's not a good idea to get between a hog and her food, we had a brief visit, ending with the promise that we could return and hang out a bit after lunch, when the pigs would be happily satiated and in the mood for naps. I would return with Bill, who knew the ropes and was designated to keep me safe within the bounds of pig-relating reason. Ha! Little did I know that Bill would cuddle up to the one aggressive pig for a wee nap, and live to tell the tale.

For the time being, we passed through the pig barn and continued on to meet a few cows out on a hillside where the grass was still green. Here Tara turned me over to Bill for the continuation of my visit, and she headed back to her desk. One of the cows we met was Frank, a now famous bovine. Frank had run away from a street-front slaughterhouse in the Bronx. He ran and ran, down 160th Street and across a college campus, where finally he was darted, tied up, and hauled to an animal control center. But because Frank made the news, Jon Stewart and his wife Tracey learned of him, rescued him, and got him to Farm Sanctuary. If Frank hadn't run for it, we'd never have heard of him. It is hard to know that the cows somehow know about slaughter and that they are afraid, that they don't want to get in the truck, and perhaps worse, that they don't want their family and friends to get on that truck and to go where it's going. How did Frank know what was coming? Why was he so desperate and determined to get away? Frank's story reminded me of moments in my own journey of curiosity, and love, and conversion, in the domain of human-animal relations, a journey that is far from over and one that has had—and doubtless will have again—some very challenging moments.

Over almost two decades, when I visited my family in England, I stayed in their house across the street from a cattle farm, a beautiful idyllic place, the British version of Farm Sanctuary, with grassy green hills and stone barns, except that here the Angus cows were intended for slaughter, not for rescue. The idyllic setting made this fact easy to ignore. Occasionally we'd hear the cows mooing, and I sometimes went over to moo at the fence and to receive a cow kiss. It was in fact one of these cows who started me on the path to become a vegetarian, when I promised her I would not eat her son. For me, beef was the first meat to go because I made that promise after seeing the love between the cow and her baby, and I will keep my promises or die trying.

One summer night in Chorleywood, I was awakened from sleep in the deep of the night by the cows across the lane crying loudly. It went on and on, literally for hours. I had never heard this in all my years of staying at that house, and I could not imagine the cause. I wondered if one of them had died. And if cows grieved? Can a lumbering bovine grieve? Theoretically, of course, I know the answer, but hearing cow grief was awful. At eight in the morning, I didn't need to wonder about the cause because the transport truck was sitting there idling while being loaded with crying cows. The cows going to slaughter and the cows staying at the farm had been separated the night before so that those tagged for slaughter were ready to go when the truck arrived. Perhaps also so that the grief of cows did not interfere with the human work of making meat, and making money.

What did the cows know? To be honest, of course, I don't want to know what those cows knew, or that they were suffering intense psychic pain caused by separation of mothers and children. But I do know it, unavoidably now. I know it from the direct experience of my senses that night and morning in Chorleywood. I heard that horrible long cow lament with my own ears. I can assure you that I'd have chosen not to hear it. But you might think me perhaps sentimental in my interpretation of the crying cows. To be honest, I might think me sentimental, too, but for the knowledge that I have of neuroscience, infant attachment in animals, and the fundamental affective/

instinctive systems that we share. Indeed, grief, a kind of primary horrible sadness, happens in mammals when we are separated from our close social connections, from our families and our close friends. And neuroscientists have demonstrated that the psychic pain of the grief system is tied to and evolved from the physical pain systems of the body. Grief is a matter of life and death in all mammals, and those cows were in the very real pain of grief.

Affective neuroscientist Jaak Panksepp was one of a handful whose rigorous research created the cognitive revolution in the study of humans and other mammals. Over almost forty years of work, Panksepp located and described seven affective biological systems in mammals composed of specific brain regions, neurochemicals, and primary behaviors that most of us would label with the less technical term *instinctive response.*[11] There are fundamental motivations in living mammalian lives that we share across species. Panksepp's seven affective systems are among them. Each of the seven systems is associated with particular brain regions, most often regions deep in the mammalian brain and considered to be evolutionarily old, and with particular chemical messengers. These in turn are associated with particular behavioral responses. The behavioral response is the part that happens in our bodies before a feeling is named or an action taken: for example, the tightening of the jaw and the clenching of the hands that happen before we even think "I'm so mad!" and long before we might threaten, swear, or throw a punch. The brain structures and chemicals and their concurrent physical expression in animal bodies are amazingly consistent across species. Therefore, as different as we are, we can know a great deal about what it is like to be a very different other, because we share some fundamental life experiences, in spite of living in very different kinds of bodies.

The seven biological systems named by Panksepp are *seeking*, or the impulse to explore; *fear*, the flight instinct; *rage*, the fight instinct; *play*, which may seem self-evident but has a very specific set of descriptors and is especially important in the social and cognitive development of young animals; *care*, the experience of bonding with another and supporting their well-being; *lust*, the desire to mate or share sexual gratification; and a kind of simple and direct *grief* or even sadness, the consequence of separation from significant others.[12] These seven systems shared by all mammals give us some common ground.

In their chapter on the primary affect of grief, "Born to Cry," Jaak Panksepp and Lucy Biven describe the grief response as "the dark side of our capacity for love and play."[13] If the soothing flow of oxytocin causes us to feel the joy of trusting social relationships, offering us the biochemical carrot that draws us close to one another, then grief might be seen as the biochemical stick that punishes us for separations in these same social bonds. The oxytocin bonding system was found to use the same brain centers for pleasure and stimulation that cause addiction to pleasurable substances and experiences. So it follows that deprivation of those pleasurable substances and experiences results in the pain of withdrawal. These fundamental affects are expressions of evolved biological values, experiences built into our very bodies to promote bonding and relationship and to prohibit separation.

Biological values? Yes. Panksepp and leading affective neuroscientist Antonio Damasio and pioneer social neuroscientist John Cacioppo all agree that these biologically fundamental affects serve to keep us connected by making the experience of connection very rewarding, and by making the experience of separation very punishing.[14] "Mammalian and avian kin-groups become 'addicted' to each other's company, thereby forming social bonds that allow them to live in harmonious societies."[15] Those harmonious societies make animal life possible, and the lack of such peaceful societies puts animal life at risk.

Scientists once assumed that animal social bonds served only utilitarian functions. In fact, John Watson, the psychologist whose experiments famously demonstrated stimulus-response learning in dogs, gave this advice to human parents with regard to their children: "Never hug and kiss them, never let them sit in your lap. . . . Give them a pat on the head if they have made an extraordinarily good job of a difficult task."[16] But scientists of the 1940s, '50s, and '60s demonstrated just how wrong the utilitarian concept was. In fact, human infants deprived of social care and secure bonds pine away and die.[17] But calves are not babies, and cows are not women. This is true, and this truth has a dark side to it in that the pain of grief is likely worse for them because herbivores bond immediately, much more quickly than do human mothers and babies, because "being born very mature, they can get lost from the beginning of life. Such mothers form rapid attachments to their offspring not only so they can flee predators together and find their offspring if they become lost; mothers also bond selectively in order to provide resources selectively to their own 'children.'"[18] Baby animals and their mothers bond at a rate related to when the infant or child has the capacity to become lost. For humans this happens in the second half of the first year of life. For cows, it happens at birth.

The cows at Farm Sanctuary were the rare ones, the ones we got to know. Here with the cows, as there with the pigs, there is a shadow cast by the light of animal rescue, and in that shadow lives the question about all that we don't know, and all that we don't want to know. I did not want to hear the cows crying that night in Chorleywood. I do not want to know that cows and calves routinely suffer horrendous grief as they are helpless to keep from losing one another; and evolution has arranged for that to be so painful that animals can die from the pain of grief.

As Bill and I wandered back down from the cow pasture, we returned to the pigs, several of whom had wandered out to sunny spots in the outdoor pens. Only one or two remained inside. A huge black-and-white boar was kicking up straw and making a nest inside the barn. Bill said he seemed to recall this boar was named Sebastian. We called the pigs to the pen rails, petted them, talked to them. And when they seemed safe enough, we climbed into the pen with them, culminating in a cuddle nap between Bill and Sebastian, photographed in its telling beauty by me. Later we learned that Sebastian is given to aggression when aggrieved, so it was a very good thing for us that he was satiated, soothed, and sleepy. Both Bill and I can testify that Sebastian is also given to cuddle naps when the occasion calls for such.

You may be one of the thousands of people described by Jane Goodall, a person who eats meat and loves animals, when she and Marc Bekoff note the paradox of doing—or being—these two things at once.[19] It's a knotty place to be, and it's easy to feel ashamed and to want to run from feeling ashamed. I can only speak for myself and from my experience, and I think the most important thing we can do is to not turn away from our discomfort, but admit that there are things we don't want to know. Just to let that be in our awareness is powerful. I think we must forgive ourselves because the train we are riding took a wrong turn somewhere long ago, and we never saw that moment. I think there's wisdom in my mother's way of allowing ourselves *to want* to want to change, even when we don't yet want to change. And when we do want to change, there are lots of baby steps we can take, beginning with knowing where our meat comes from and choosing meat that has been humanely raised and slaughtered. That is what I did for quite some time, until I faced up to the fact that, yes, I could locate humane farms where I lived, but when I traveled, the meat that I was accustomed to eating was not available in airports and the restaurants that serve travelers, and I was eating meat anyway, bad meat, badly got. It was best for me to stop eating meat.

The next day, Bill and I returned to sit in the sheep barn for a couple of hours. I sat there trying to absorb the experience of Farm Sanctuary. I wished I had the clean conscience that Bill has, but the contrasts were so stark and my inner eye could not quite adjust to find the defining edges of things. The sheep and goats were curious and welcoming. Skye and an older ram whose name I did not know vied for my attentions, pushing at one another, each trying to get closer to me. It was a beautifully primal moment in which I experienced the competition between these two old boys for the attentions of this one old girl. Bill went off to look for his friend Dotty the goat, who remembered him and bleated happily in his company. I photographed the two of them, as I'd photographed Bill and Sebastian, the boar. I let the questions arise inside me: What about the pigs we never met? What about the cows whose cries we never heard? These, I think, are the uncomfortable questions on our plates. But being uncomfortable will not kill us, nor even harm us, though it may cause us to change. I recalled my attempts to crack through the laser-like reasoning of Susie Coston, to get at the core of her motivation for doing this work.

"But you, Susie, why do you do this work?"

"Well, first of all, I was at a place when I came into this work where I could have easily—I could have been dead in a couple of years. I was really messed up. I had a lot of my own problems and I started working with animals and they just disappeared, my problems. And because these animals gave me this thing that I did not have inside of me, I love them and I feel like they saved me. And then that became my goal: I will now save you because I have a different power than you have. It was pigs that I had started working with, and I had never been around a pig, and they are so incredibly intelligent, and they are so appreciative when you save them. The animals live in the moment and they allow you then to start to live in the moment. And then all the other stuff that has complicated everything for you kind of disappears. So I just find that I have to stay focused on what I love and keep doing what I do. You have to do what's in front of you, or otherwise all this stuff will weigh you down to the point where you're paralyzed and you can't work. This isn't Utopia, I do see the big picture and I do feel the frustrations, but at the same time, if I let all that eat me alive I will just sit around complaining on the computer and critique everybody else. So I just give the beings that I love the most good I can, which just happen to be these animals."

What to do with unwelcome knowledge about farm animals? About anyone for that matter? I think there's tremendous wisdom in Susie Coston's approach. Do what you can for the one in front of you, avoid getting lost in enormity, look someone in the eye and love them, offer care. Then get some rest and do it again.

Of my visit to Farm Sanctuary, I have to say that the idyllic, intentional, interspecies community visions that had brought me here receded rather quickly into a much more distant and indistinct future. The meetings with

Julia and Sebastian Pigs, and Dotty and Benedict Goats, and Frank and Diane Cows, on the other hand, took on a much more defined reality. These were individuals who challenged the philosopher in me about the utility of defining what *person* means. I found Susie and Bill to be unlikely saints, with nothing plaster or pious about them yet with real stripes of saintliness in each of them. There is a lot of light in Susie's laser focus on doing what she can do and being the best possible version of herself in the doing. And there is a lot of light in Bill's arrival late in his life to clarity and purpose, at least in this domain wherein he can use the qualities that are his nature, his habits, his skills, in order to leave the world a better place than he found it. Of it all, I can say that it seems incredibly complex. Or is that merely convenience, to see that what is wrong is complicated and somehow impedes doing what is right? Does it matter if the chicken's mind does or does not think in such abstractions? The chicken loves her life with or without access to such abstractions, as do we.

Skye died recently. I cried when I read the news. I think I will never forget him, with his joyous bright eyes and boisterous head butting. I can't imagine the goat-and-sheep barn without him, though I am sure the resilient animals have reconfigured. I know as I write that Bill is giddy with packing his suitcase to return there, and I await his report, even as I don't want to know what it is like without Skye, the lead goat who became for me a person in my brief visit to Farm Sanctuary.

It's hard to know about the millions of cows and pigs and chickens in factory farms. And it's inspiring to know that somewhere some human is strapping wheels onto a lame goat so that he can tear madly and joyously in all his adapted goatiness down a grassy hill. Somewhere, human police are arresting people who abuse farm animals, and carrying those animals to people who will give them care and shelter and a chance for healing, perhaps healing themselves in the process. Somewhere a human is sitting with a lap full of genetically modified chickens with their little wings over one another.

CHAPTER 3

The Big Five

AND THE LITTLE FIVE MILLION

We were rolling along a dirt track toward the end of the day when Magnum signaled Edward, the ranger and driver, to stop. There was a collective intake of breath as the ten of us in the jeep searched the dusky bush with our eyes, cameras at the ready, looking for one of the "big five." This term coined by big game hunters refers to the most difficult animals to hunt on foot: the lion, the leopard, the elephant, the cape buffalo, and the rhinoceros. It was just dark and we were almost back to the lodge. We were ready to photograph the elephant or buffalo, the rhino, the lion, maybe even the elusive leopard as it moved through the early dark, if only Magnum could get one of them within his night spotlight.

But Edward was pointing to the ground next to the jeep as he cut his lights. We'd stopped rolling, but the animal of interest continued to roll right along beside the behemoth vehicle, a classic safari jeep with tiered seats and the tracker dangling off a seat hanging from the front guard rail. Magnum, the tracker, trained his light down toward the left front wheel. There it was, a dung beetle, one of the little five million on whom the big five and all of the rest of us depend, and certainly not one of the most savory of the little five million. There are different kinds of dung beetles, but the one thing they have in common is that they are all in the business of—you guessed it— dung! Some of them simply live in it, some of them eat it, and some of them make bridal chambers from it. This beetle turned out to be two dung beetles, male and female, moving their bridal chamber together.

Our dung beetle species was *Scarabaeus satyrus*, to my mind the species that lifts the entire world of dung beetles out of the proverbial toilet and into the starry realms of romance. This beetle labors at night, gathering and rolling balls of dung for long distances, then digging chambers in the soft ground and burying the treasured balls in the place that will soon be both bridal chamber and nursery. Not only does the beetle gather and roll balls of dung, but it does so at night, walking backwards in a perfectly straight line. Recent research has illumined the brain features that allow it to do so.[1] I think Ginger Rogers would be proud.

You might wonder how they can do that, remaining in a straight line for a long distance at night. It's a question that has puzzled scientists too. They reasoned that the dung beetles might use the light of the moon on moon-lit nights, but what about when the moon is dark? And they might use polarized sunlight for an hour or so after sunset, but what about during the middle of the night when they are quite active? Before any neuroscientists could locate the brain mechanisms that make the navigational feat possible, someone had to demonstrate what it is that actually describes the behavior of the beetles. Just what are they up to? What are they doing? It was a team of scientists performing experiments inside the Johannesburg Planetarium, controlling for sources of light, who discovered that dung beetles orient their straight lines with reference to the Milky Way. Commenting on their conclusions, the team summarized the significance of the study: "This finding represents the first convincing demonstration for the use of the starry sky for orientation in insects and provides the first documented use of the Milky Way for orientation in the animal kingdom."[2]

This fact was something I had learned between my first and second trips to Kapama Private Game Reserve, just outside of Kruger National Park in South Africa. As Edward spoke of the nutrient-rich courtship of these two beetles and extolled their virtues as recyclers, I recalled that I had felt bored and frustrated when, on my first trip, a ranger had wasted my precious time in the bush pointing to and talking about dung beetles. This time I was enthralled. Not only had I learned about the Milky Way navigation, but I had read Bernd Heinrich's *Life Everlasting: The Animal Way of Death*, where he describes the intricacies of recycling that are at the heart of every ecosystem, wherein nothing is wasted and one thing morphs into another, life without end. Heinrich describes elephant dung in some detail: "Elephant dung is very coarse compared to any other dung I have ever seen, smelled, or felt. That is because elephants don't eat just young juicy grass shoots; they eat whole bushes."[3] He adds, "If you go out after dark where a herd of elephants has recently come through . . . you will hear, shortly before the deep cough-

ing roars of the lions begin, a low humming noise. It's the sound of dung beetles barreling in—hundreds and then thousands of them—all making a bee-
line to your dung pile."[4]

This is the thing that the rangers and trackers want us to know, that the
bushes and the elephants and the dung beetles are one interdependent
thing. Many rangers and trackers have told me they also first came here for
the excitement of the big five, and that they deepened into love of the little five million. They are really excited to show us one aspect of the complex relations of mutual dependence that make the African bush a living
ecosystem, a life illumined by the sun of elephant daylight and by the deep
dung-beetle light of the Milky Way at night. The big five and the little five
million are continuous with one another.

My previous visit to Kapama had been by happenstance rather than planning, having stumbled into the fulfillment of a childhood dream. At that
time, I was teaching college classes on a ship that was circumnavigating
the world, and I was assigned to supervise sixty undergraduates on a safari
field trip to Kapama, a big and growing conservation project about which I
knew nothing. I had learned upon arrival in the company of jeep after jeep of
students that the facility included a 24/7 bar and that the drinking age was
eighteen. Needless to say, I had been distracted from the animals by my duties, almost to the point of breaking my heart, and I had promised myself
that I would return.

When Diane and I arrived at River Lodge, this time without the respon-
sibilities, we got out at the safari staging area where jeeps pull up to ramps
and humans pile in and out of them. This feature is reminiscent of a theme
park, which I suppose Kapama is, in some sense. It is also practical, as many
visitors would otherwise have difficulty hoisting themselves up into the jeep.
Walking along the raised boardwalks under thatched roofs, I felt confirmed
in my memories of the place, and I felt the pleasure of seeing Diane—who
had agreed to accompany me as a kind of research assistant—see Kapama
for the first time. When we arrived at the River Lodge *boma*, with its hide
rugs, thick pillars, thatched roof, huge central stone fireplace and, yes, bar,
we were greeted by Liezel Holmes, the head ranger and safari manager. Liezel is petite with wiry strength and a certain gravitas that doesn't necessarily come with the ranger uniform of short-sleeved khaki shirt and shorts, ev-
idently ironed with starch. I was tickled to introduce Diane to Liezel, whom
I remembered easily, having had a very vivid experience in her company.

Eight years earlier I had been sitting next to Liezel in the front of a safari jeep when we came upon a bull elephant trumpeting and stomping and

moving onto the dirt road in a furious display. Liezel quickly put the jeep into reverse as we caught sight of another jeep approaching the elephant from the opposite direction. I watched first in excitement and then in horror as the elephant swung its trunk over the tracker who crouched down in his seat suspended from the front of the other jeep, his hat flying through the air, and I remember a woman in the second row dropping to the jeep's floor without waiting to be asked. We were relieved to see that the elephant had removed the tracker's hat but left his head in place. And I remember Liezel, with steam piping out her ears and through her clenched teeth, telling me that the ranger had come too close to the elephant, and that his ego would be tamed by a week of bush-clearing work. Obviously he wasn't ready to drive the jeep. Obviously she was the boss.

When we sat down to a formal interview, I asked why she was in this work, noting that she had been at it for a long time. "It's the animals!" exclaimed Liezel. "Ask any one of these rangers or trackers, even the staff. They do it for the animals." She went on to describe her own knowledge deepening with experience, and her own love for the bush growing in complexity as she experienced more of it, especially as she came to see the five million. She described going through a phase of laying nets in the grass and under trees, just to scoop up a catch of insects, to look at them closely, to learn who was there and what they were doing. And she told me about her joy on the day she finally saw a pangolin, that she came ripping home bursting with joy that evening and the rangers drank a toast to her and her pangolin sighting.

In the course of that interview with Liezel, I learned that Kapama was approximately 14,000 hectares (about 34,500 acres), including four distinct lodges, along with Camp Jabulani and the Hoedspruit Endangered Species Centre, all of it owned and operated by the Roode family. The luxurious lodges had a rather astonishing 80 percent occupancy rate and Kapama offered, for a price, a variety of experiences, from romantic sleep-outs in the bush to spa services, private game drives, bush walks, and (once upon a time) elephant-back safaris. Even as the whole project depends on the satisfaction of high-paying visitors, the work of Kapama also includes community outreach, especially leading local anti-poaching efforts, including working with local law enforcement and providing anti-poaching training, and offering education and support for abandoned children. This was especially touching for me, the outreach to abandoned human children, in tandem with the care for abandoned elephant and rhino children, all of whose futures likely depend on one another.

All of the Kapama programs, the four lodges, the Endangered Species

Centre, Camp Jabulani, and the community outreach programs began as a seed more than sixty years ago when then-six-year-old Lente Roode was given a cheetah cub whose mother had been shot by a neighboring farmer. It was common then for farmers to shoot and kill troublesome predators, who of course became more troublesome as their range was reduced by human farming. Lente named the cub Sebeka, and the two became inseparable. Years later, Lente and her newlywed husband Johann bought property next to her father's farm, intending a cattle ranch. But the predators were challenging and the couple responded by changing their holdings from a cattle ranch to a game preserve, which almost necessarily evolved into a conservation project. Johann was a hunter, and Lente freely admits that it was an ongoing area of disagreement, that she was never supportive of hunting the animals. When her father died, Lente inherited the family farm. Later, Johann died suddenly and unexpectedly, leaving Lente to set the direction for the twelve hectares that evolved into what Kapama is today. Lente Roode has now received a Lifetime Conservation Achiever Award from the Wildlife and Environmental Society of South Africa for her many contributions to conservation.[5] It is clearly her passion and she says that it started with Sebeka, the orphaned cheetah. "My problem," she once told a reporter, "is humans."[6]

Happy as I was to find the River Lodge boma and staff as I remembered them, I soon noticed that the game drives were, well, different. I noted this apologetically to Diane, whom I had prepared to see the bush teeming with grazing animals. I'd told her we might even see lions if we watched carefully, remembering the sound of a sharp intake of my own breath and the way the little hairs on my neck would stand up when I caught sight of the golden eyes of the golden lioness, crouched in the golden grass, the lioness who had been watching unaware me with calm appraisal. At first I thought maybe the lack of abundant grazing animals was just the luck of the draw, a result of things like the time of day or the particular location. But it was the drought, as Edward told me, the same drought that I introduced in the first chapter of this work.

And there was a particular twist to it: we saw many predators. Lions seemed to be everywhere, snacking on carcasses, sunning, sleeping, strolling along, relaxed and happy. We stopped to watch a group of lions tearing at a buffalo that had been killed the night before. Far from being wary, they seemed to barely register that we were watching them. We even heard the mating call of a female leopard, that soft growl that penetrates the bush, like a human singer holding a note both soft and loud at the same time, a whisper that can be heard from a distance. We tracked that leopard for almost half an hour, watching her move through the bush until she hopped into a tree the way a domestic cat hops onto the kitchen table. Leopard sightings are rare. We were to have two extended sightings over the course of our time here. But while the predators seemed many, the grazing animals were not nearly as abundant.

The last time I'd been here, also in summer, the grazers were thick, the primates—baboons and monkeys—were many and loud, and I was struck by how well they got along, different species sharing the same space contentedly, with only the occasional startle or argument. Zebras, warthogs, giraffes, wildebeest, nyalas, gazelles, all in approximately the same location, with monkeys in the trees. Even with lions nearby, the grazers had not twitched with fear or jumped away from each other as too many years of king-of-the-jungle descriptions of African wildlife had led me to expect.

But that had been a rainy year, and this was a drought year. The grazers flourish in wet years because there is plenty for them to eat, and they can make many babies who thrive because the adults are thriving and the world around them is sustaining for them. The wet years are not as good for the predators because the grazers are stronger and faster and they hang together in protective clumps. That was what I'd seen before, many relaxed grazers and a few wary predators, quiet and difficult to spot, and working hard for their living. What I was seeing now was the opposite situation: scarce water, grazers few and weak, easy prey for the many thriving predators. Even the dagga boys, the old bachelor water buffalo who hang out together, were not hanging out together but scattered and weak, and I think I saw as many buffalo remains as living buffalo.

In the course of my interview with Liezel, I brought up the lack of grazing animals and asked her if my perception was correct, wondering if she would connect current conditions to global climate change. She confirmed that the grazing animals had had a huge die-off and were struggling, and offered that the root cause was the drought, that the cool weather and the water never came that year. And she seemed to almost fold up when she offered that this kind of drought was not normal, not par for any course anyone here had heard of, but, rather, something new. "Global warming," she mumbled bleakly. But then she drew herself up, offering a slim but real hope: "Unless humans decide to do something from their side—. Years ago you didn't hear about conservation, now boards are going up and you just see conservation signs all over, which is fantastic." I have seen over and over this sense among conservation and animal-rescue workers, the wavering in the face of seemingly overwhelming odds, and then the grasping more tightly on hope. What else? Each faces into the work that he or she feels compelled to do instead of admitting defeat from the outset.

Saddened as I was by the effect of the drought on the grazing animals, and much as I longed for the throbbing sense of life they give to the veldt, I came to see that my fascination with them and with their predators had distracted me from many other kinds of animals. As the dusk deepened, Magnum, who had seemed to me subdued, suddenly came to life. His flashlight combed the bush and soon a bush baby, a small primate with huge fixed eyes and huge ears adapted for night life, turned its head toward us, its large golden eyes dominating its appearance. We saw a jackal, a wild canid of a species that mates for life in an affectionate bonded pairing, as many species do. This jackal had two puppies tussling together nearby. This was to be one of the great pleasures of our time in the bush, the abundance of animal babies of all kinds to be seen in late spring and early summer. As the dark deepened that evening, Magnum spotted an animal I'd longed to see, again as the tracker's flashlight caught the reflection of large nocturnal eyes, this time of a small-spotted genet, reminiscent of a cat in its body size and movement, but also of a raccoon, with its very long ringed tail, and of a fox, with its pointed little face. Both bush babies and genets are sometimes kept as pets, though increasingly they are illegal pets. And, for the would-be genet owner, there is a sobering warning: plan to keep the animal for its life because genets cannot tolerate abandonment, but become severely mentally ill if separated from their initial owner. I wonder how many people had that experience before it became common knowledge. I wonder how many genets went crazy from the grief of abandonment. This genet did not hang around to bond with us but cat-walked off into the darkness, trailing its double-sized raccoon tail, and we headed back to River Lodge for dinner.

We were informed as we departed the jeep that we should meet back at the ramp by 5:45 a.m. the next morning, ungodly early but before the heat and, even more importantly, as the animals awakened, most of them being particularly active at daybreak and early evening (crepuscular to some degree), at least during the summer. Edward and Magnum kept us in the jeep as they swept their flashlights carefully around the ramp. They had told us to wear boots, not sandals, on safari drives. To reinforce the request that we wear sturdy footgear, we heard about the ranger who, coming back to the boma wearing sandals after driving from one lodge to the another, was the victim of a black mamba in the parking lot. As the *National Geographic* delicately describes the snake whose very name strikes fear, "Black mambas are fast, nervous, lethally venomous, and when threatened, highly aggressive."[7]

Black mambas are actually much more often olive, gray, or brown, grow up to fourteen feet in length, and are among the fastest snakes in the world. It's fine with me that I have never seen one. And I always wear good boots in the wild, having lived for many years in rattlesnake country.

It is surprising how full and how tiring a day can be that revolves around four to five hours of sitting in a jeep. Diane had no trouble rising in time to be at the jeep at 5:45. Indeed, she'd been awake for an hour, quietly twiddling her thumbs and waiting for me to make a sign of life so that she could bounce. We had coffee that first morning on the veranda at a few minutes before five, and it became our practice for the duration of our time in the bush to rise in the quiet dark and converse in soft tones as the day gradually came on and the volume went up around us. Then we checked out the various snacks on offer to tide us over until breakfast and headed to the loading ramps. We also made a habit of sitting in the high back seat of the jeep.

On that first morning drive we started out chilly, wrapping ourselves in fleece shirts and blankets, but by our return, still early morning, things had heated up, and the blankets had been dumped in the jeep. I took it as a lesson in climate wisdom and headed for the gift shop for a long-sleeved lightweight solar-protection shirt and summer bush hat, neither of which I had brought with me on this multileg journey. I quickly found exactly what I needed, in light beige, the recommended color both for camouflage and for avoiding attracting mosquitoes, both items bearing the Kapama logo, and both serving me well to this day on my farm in southern Italy.

The last time I was in this gift shop, I'd bought a beautiful and well-made bracelet of elephant-tail hair, wrapped in silver. There was nothing like that in the jewelry case now, though there was plenty of silver jewelry in the form of various animals. Humans have perhaps forever loved to wear representations of their totem animals, but there were no actual body parts, not even a hair, now included in the jewelry. I would not buy that bracelet now, even though I know it was made from a hair collected from the ground, even though I had enjoyed the feeling of some kind of intimacy with the animal I so love. Now I understand that it sends a message that things made from animal body parts are cool. It's a message I don't want to send, and it's a message Kapama has decided not to send. I have noticed that realizing dreams typically includes a loss of innocence, a bursting of the bubble, even as the dream is fulfilled and fulfilling. Loving animals means wanting to get near them and getting near them isn't always good for them, depending on how you do it. Before leaving River Lodge this time, I returned to buy gifts of fabric, beautiful African fabric.

Soon after our morning drive began, Edward stopped and backed up the jeep, then hopped out and took a leaf from a tree. Oh man, a leaf is not a leopard. I could feel that thought in the collective sigh. But it wasn't a leaf. It was a chameleon. Edward showed us how it had already begun to change colors, and offered to let us hold it. I was eager and very pleased with the brilliant green and bulging eyes and tail wrapped like a fern bud. The chameleon was passed around but only Diane and I and one other adventurer were up to a close encounter. After we had held the creature, watching him change colors and listening to Edward's description, Edward put the chameleon onto the trunk of the tree. We watched him turn brown with a bark-like pattern within a minute. I took the quiet occasion to ask Edward if he had a favorite season at Kapama, and he replied brightly that it was now, late spring and early summer, because you get to see so many behaviors. I found it funny the way everyone at Kapama seemed to use that word, "behaviors," because animals are always behaving one way or another, since even sleeping is a behavior, and humans are behaving too, but we don't typically report delight in seeing an array of human behaviors.

Soon enough we came upon two young male giraffes competing for dominance. Edward signaled this as an example of the kind of behavior you get to see at this time of year. It was an impressive fight, primarily because of the level of tension evident in the two males, who stood next to each other, vibrating, watching out of the corners of their eyes for long minutes, until one of them would swing his powerful neck and belt the other one. The sound of the contact is shocking. The tension as minutes go by in vibrating stillness and the intensity of the fight are equally palpable, and the quality of the animal's life depends on winning this battle: one of them will control the territory and the other will live warily or move on. We would see these two young males carry on this competition over two entire days. How exhausting it must be, and what a drive to win!

This male competition for control and access to food, to range, to females, is classically described in terms of evolution. The strongest, smartest, best one wins, and he makes better babies. The rangers and trackers describe every animal behavior in terms of adaptation, bringing home the point that most of life comes down to eating and mating, with temperature control perhaps coming in at a distant third. While these filters for observing the animals are accurate, I think they are not sufficient. These are not just eating and mating machines, at least not any more than we are. They live complex social lives and they have individual temperaments and interests, as we do. They have a world of thoughts and feelings inside and a world to which they relate outside. And there are moments when the rangers and trackers break out of the "behaviors" explanatory mode in bouts of excitement, as they describe some event that gave them a sense of real encounter with an Other.

I asked Liezel about this, about the idea many people have that humans are superior to other animals, and she responded in a flash, "*They* are superior!" She said that we are weak and dependent on tools and they, the other animals, live from their wits and the capacities of their bodies. She advised me that if I ever doubted it, I need only be charged by a lion while walking through the bush, even with a rifle in hand. I asked her if that had happened to her. "Oh yes!" she replied. The lion, she says, just wants you to back up, to get out of his territory, away from his pride. You can't run, that's what prey do. You can't fight him; you'll die. So you obey with dignity, careful to signal that you are not a weakling to be eaten as a snack. You back away quietly, your head held high meeting the lion's gaze, not screaming or crying as you might very well be prompted to do. And not wetting your pants either because that, too, is a sign of weakness. To my great relief, we wrapped up our second drive without seeing a kill, in spite of the large numbers of predators and the weakness of the prey. I tend to be sympathetic to the prey, and I have been reprimanded for this by wildlife biologists who point out to me how very hard is the predator's life, always on edge for the next meal, and often not getting a meal, going to sleep hungry.

Our dining room boma was a huge area under a thatched roof, with multiple buffet tables, and with tables and chairs for diners under the thatch but also outside and open to the elements. It was surrounded by a dense cover of trees and shrubs, squirming with monkeys and more monkeys, crafty monkeys who want food. To sit outside is to court them, intentionally or inadvertently. Of course, feeding the monkeys is not a good idea; it is not allowed and the staff makes every effort to educate anyone with the slightest inclination to use food to court the monkeys. Having had my cheek and lip neatly split open by a monkey (whom I was not feeding!) in India, I exercised caution with these African monkeys. Still, the monkey viewing was compelling, particularly because of the season, with lots of baby monkeys hanging on to their moms and playing with each other's tails, shrieking, laughing, and crying, as kids at play do, until a staff member appeared with a broom to shoo them and to remind us and them of the rules.

The time between breakfast and the evening safari drive is when you can do other things, like take a guided bush walk, or have a massage or a nap, or a visit to the Endangered Species Centre, or a few drinks in the bar, or visit the program for schoolchildren. There's something for everyone. We chose an elephant-back safari for the first day, a guided bush walk the next day, and a visit to the Endangered Species Centre the next. The intention to satisfy the desires of visitors is no doubt what keeps the place fully occupied, and it also forms an interesting juxtaposition that exposes us, the visitors, the tourists. We want to experience nature while also looking great for the photos we will post for friends. We want to support conservation. We want to see the big five. We want to take good photos and tell exciting stories. Some of us want to witness the kill. We want to rest and renew and have a relaxing massage. I note that we are a complicated species and that the other animals pay dearly for that fact. Kapama and many other animal-rescue and conservation programs try to bend our weaknesses, our follies, into strengths for the greater good. It seems a somewhat precarious balancing act, a kind of eating the cake in order to have it. And yet here it is, here are the animals, living in relative safety and peace. And here are the humans, acquiring an education, certainly, but also perhaps having their hearts opened, having their souls framed by the vivid and uncontrived life of the bushveldt, and perhaps seeing themselves differently in the context of that frame.

The afternoon drive at four is hot, and I am grateful for my light shirt and shady hat. We've been advised to bring some money, just in case we want to buy a drink out there on our rest stop. Some of my favorite moments were those rest stops, not so much for the drinks, though anything cold is exquisite, but for the opportunity to sneak away on foot on my own for a bit of biorelief, to see close up the scorpion tracks at the base of the termite mounds, and the warthog burrows, and the little flowers blooming underfoot. This particular assemblage is one that the rangers like to talk about, too, as they like to talk of dung beetles and elephants. This is an assemblage of termites in the huge reddish termite mounds that they construct, and here are warthogs in dens at the base of the soft soiled mound, and now scorpions are moving in, all making tracks in the crumbly turf around the mound. From this point of view, it is not one thing that matters, not the termite, nor the warthog, just as it's not the elephant or the dung beetle. It is the relationships among the beings that make this a place of teeming joyous life. Before exposing any flesh, I carefully scan for the black mamba and the little scorpion.

On this drive we had what was for us a first viewing of a mongoose, emerging from its den in the late afternoon. I squealed in delight, "My first mongoose," and Diane gave me the look that means *notice yourself*, because squealing is not a good behavior for attracting wild animals or for pleasing fellow travelers. Yes, my enthusiasm had gotten away with me, but can you blame me? The little mongoose is a darling but fierce mammal, one who can take on a black mamba and win.

When I interviewed Liezel, I'd asked her about rhinos because, after three drives, we'd not yet seen them. "*Yes!* We have rhinos. You will definitely see rhinos," she enthused. "Oh, fantastic!" I responded. "How many rhinos do you have?" Liezel just kind of contracted and replied that she couldn't say because that is information, and any information can be used to poach the beleaguered rhinos. I was taken aback that such an innocent question could provoke caution. When we actually saw the rhinos, many of them, I was so much more appreciative of their precious existence. They grazed on a great plain covered with shrubs and surrounded by trees, seeming to move in one general direction but in loose groups, in friendship pairs, occasionally a mother walking behind her baby, giving delicate directions with her horn. These giants don't see well at all, making me wonder why evolution bothered with eyes for them. By hearing and smelling is how they get around and how they communicate with each other. It makes them dreadfully vulnerable to poachers who dart them, neither seen nor smelled, from a distance, even from helicopters. We spent quite a while among them, just sitting in the jeep, talking and hanging out. I got kind of fascinated with their gigantic three-toed feet and tickled by their grace in using them. Edward pointed to two rhinos who had walked a long way together. He told us they were two females, and best friends. A rhino girl with a BFF? Who knew? Despite their reputation as being aggressive animals, none threatened, much less charged us. The danger goes the other way.

Halfway through our time at Kapama, we moved from River Lodge to Buffalo Lodge, a much smaller and more intimate setting, composed of ten large tents set on wooden platforms in a dense canopy of trees. The tents are beautifully furnished in the style of nineteenth-century European safaris, only with the addition of air-conditioning and Wi-Fi. Imagine taking your hot bath in a deep tub next to a glass wall looking into the canopy of trees, with animals grazing down below. It is quietly spectacular. And at night you hear the animals, monkeys scrabbling around on the decks, elephants rumbling and pulling up small acacia trees.

When we arrived at Buffalo Lodge, we were greeted by a man holding a silver tray of hot washcloths. For each of us, he lifted and offered one with silver tongs. I noticed that he was reserved, in the manner of someone consciously withholding himself. Of course that very fact sparked my curiosity. Normally my response would be to prod to find out why the reserve, but on this occasion I landed on respect and decided to let it unfold. The lodge manager was there to greet us and to provide an orientation to our new digs. I knew that she and Liezel had talked about our transfer, so I teased her that I had gotten a complimentary bottle of wine at River Lodge, and now the competition was on. Then she introduced us to our ranger, Aneen, a lively young woman who spoke in the soft Afrikaans accent, and to our tracker, Fossie, the man who had with the best of manners given us the warm cloth but nothing of himself.

After lunch, we prepared for our first drive with Aneen and Fossie. When we arrived at the ramp, we learned that our game drive would be private, that the two of us and the two of them concluded the head count. We could not believe our good fortune. As our time unfolded, we came to understand that every one of our game drives over the next three days would be a private drive with these two highly skilled and delightful people. There would be no pressure to make sure we saw the big five, because we'd already seen them and there was no one else to please. So instead of pursuing any particular experience, the four of us would be wandering around in the bush exploring together. Each of us had her and his own deep love of the natural world as a core element of our selves, so we had a fundamental liking of each other. Both of them were into appreciating the whole integrated thing. These facts determined the quality of our time together; it was nothing short of beautiful.

As was common here, Aneen slipped into Afrikaans when she needed to have a private conversation with another ranger or staff member. Fossie was an amazingly skilled tracker because he had vast knowledge of the animals, but also of the plants, the weather, the soil, and how they all interacted. We followed fresh lion tracks one morning for thirty minutes, with Fossie stopping to look, to sniff, to get out and test the ground in order to interpret the age and weather conditions of tracks. Finally, he felt we were very close. He and Aneen got out of the jeep, she with rifle in hand. I started out of my seat behind them but they would not allow us to be on foot on this occasion. They soon came upon the lions and returned to drive us close enough to see them resting at a small waterfall and pool.

As skilled as he was with knowing the animals and how to find them, Fossie was equally interested in the plants, even the soil conditions. He stopped frequently to tell us about the relationship of a particular plant to a particular animal, or about the medicinal qualities of a plant or why the tracks look one way in this kind of soil and another way in different soil. Aneen told us that she had decided that she would specialize in birds, that she would be the one the other rangers turned to for answers about the birds. And know them she did. She could imitate the call of any bird we met, and she often stopped the jeep when she heard them in the distance, to tell us about them or to call to them. "Whad-up? Whad-up?" she asked the hornbills.

Early on, Aneen stopped to point out a bird high in the tree overhead. I didn't yet know that birds were her major, so I asked Fossie, the official tracker, "What does he say?" He looked at me with a puzzled expression and I clarified my question: "What sound does that bird make?" Fossie grinned and spoke bird. Fossie went on to show us the complexities of termite mounds, warm biomasses whose bases were filled with nesting creatures, from scorpions with their tiny railroad tracks to opportunistic warthogs, who simply took over dens created by mongoose. Yeah, I had heard about that, but this time we walked over, and I could see the various tracks clearly and could feel something of the relationships of one thing to another.

Near the start of one morning drive, Fossie signaled for Aneen to back up the jeep, then he hopped off and darted into the bush just off the dirt road, returning with something in his hand. He showed me a stone and asked if I knew what it was. I grinned as I took the tear-shaped stone into the grip for which it was made. I was delighted to feel this old hand axe made before industrial times for cracking bones and nuts. I recognized it because I'm drawn to old Neolithic sites and have spent many hours crawling around in them and have held hand axes before. Fossie demonstrated its use in the air, and told us it was for cracking bones, then turned to replace the stone on the ground, but then he turned back again and placed it in my hands. I certainly wanted it, but I asked, "Are you sure?" I had acquired a feeling about Fossie, that he was associated with the sacred in his community, so I thought he would understand my desire to be respectful of where old things do and do not belong. He assured me that because of my research and my work, it was an appropriate gift, that I really could keep it. Aneen assented. I accepted.

Fossie and Aneen and Diane all knew much more than I about flowers and plants. Diane was frequently able to draw comparisons to wild plants in the western United States. We stopped for marula trees, beloved of many, but especially of elephants, for their bark and fruit and for the water that can be got from their roots, the water in the roots that those sensitive elephant feet can hear in the ground. The marula is popular with humans, too, because it can reputedly determine the sex of human babies, each tree being either male or female. We stopped for a plant with aloe-like slime, which we discovered with enthusiasm because we keep aloe plants that look much

different from these but have the same slippery juice and the same properties for healing burns and cleansing guts, which fact we shared enthusiastically. One time Fossie signaled to stop the jeep, then reached under the
seat for a machete and began to hack the ground beneath a smallish plant.
Soon he had excavated a huge bulb, a bulb that overflowed his two hands.
He was obviously delighted with his find. He said he would cut and dry it
and it would be good medicine for a variety of ailments, for humans and for
animals.

One evening drive, Aneen, who had planned a treat for us and who knew
where she was going and why, made the occasion for my favorite photo of
the entire trip. It shows blackish-blue round objects with fat feet, in diffuse
moonlight: hippos emerging from the water for their evening graze. We'd
seen a few earlier and, more to the point, heard them, fighting and flash-
ing teeth. Now we saw a great number of them emerge from the water to the
grasses, creating for us another of those "Who knew?" moments, as we saw
the secret life of hippos unfold in beautifully indistinct visuals, contributing
to the feeling that we were magically incorporated into a different world.

The next morning, our last, it was crocodiles, several of them. Again I was
relieved to neither see nor be the kill. That was also the morning on which
Fossie noticed a rock protruding from the earth, from which we all together,
under his guidance and with the assistance of his strength, created a birdbath at the base of a tree. This shared creation served as a kind of ritual act
to psychologically cast in stone the bonds we had created with each other.
For Diane and me it was a beautiful and aching farewell to Aneen and Fossie
and to Kapama.

Leaving Kapama in the late morning of our last day, Diane and I were
quiet as we surveyed the dusty roads and tall grasses of early summer, trying to imprint them on our minds and hearts, watching for the animal surprise that happens so easily when you're not looking. All of this is a long way
from home and we each carried a new sense of its vulnerability, a new grasp
of the fact that it really might not be here in this magical way in another decade, another century. We had made note of the dreadful effects of one long
drought. We were a bit subdued, perched in our accustomed spot, the high
back seat of the safari jeep, even as we glanced occasionally at our bouncing
luggage. When I heard an unusual cat sound, a low call with a longish rum-
ble on the end, not a roar, not a purr, but something more plaintive, I said
with excitement to William, our assigned airport driver, "Did you hear that?
I think it was from that lioness. Look! She's walking away from the group."

"Oh, she's looking for love," said William. "She's flirting, calling for a male to mate with her. Look for the male." The lioness continued to walk away from the group at the watering hole, a seemingly strange way to engage one of them for mating. But then we saw the male lying at the watering hole as he raised his heavy dark mane, stood up, and put on a determined stride. He walked about a hundred yards behind her, even as she crossed the road and walked along the shore of another watering hole, occasionally looking back over her shoulder to encourage the determined dusty male. The hot sun was high in the sky and he seemed to hesitate, to want to stop for a rest or a drink of water. But she seemed to promise that it would be worthwhile. He was unable to not obey the summons. He was still following behind her as we drove on. They would likely spend a day or two together copulating briefly and often, and napping and cuddling.

Somehow this cooperative flirting, initiated by the lioness, was not what I expected of lion mating. The narrative I have collected over and over again over the course of several decades is that male animals compete with each other for females and, after vanquishing lesser males, dominantly claim the females of their choice. I expected that all of this would be especially true of the king of the jungle. But that was not what our eyes, ears, or the rest of our animal bodies witnessed that hot December morning. What we saw was a cooperative dance of lust between two animals, initiated by the female, pursued by the male, carried out in the challenging circumstance of the heat of midday. It was an enlivening end to our time in the bush, witnessing the sexy lions perform their invitation and response to each other, suggesting not the end of everything—our visit, the thick pulse of animal life in the bush, the Earth herself—but renewal, the knowing of lust as we know it in our own adult mammal bodies, knowing, too, the promise of another generation that is given in the depths of that lust.

The two lions we saw that morning were responding to the circuitry for lust buried deep in their mammalian brains, the urge to merge that all mammals share, an urge that is closely related to the seeking system in the brain and also to the care or social-bonding system. But although lust shares some space with these other two systems, it has its own distinct niche in the brains, bodies, and behaviors of animals. When neuroscientist Jaak Panksepp describes the seven affective systems in the brain, of which lust is one, he makes clear that what he means is a location in the brain that is associated with specific neurochemicals and behaviors.[8] When animals share these physical systems and related behaviors, we can strongly infer that all animals share similar internal experiences of feeling.

Let's pause on this note for a moment: internal experience of feeling is exactly what scientists for several generations believed that only humans possessed, that the other animals did not have internal experience, but were literally more like machines programmed to operate in particular ways and unable to operate in other ways. It was a claim that originated with Descartes, once known as the father of modern science, and with his mechanistic philosophy. The world outside of humanity is not sentient, Descartes said, but a great machine in which one thing affects another in necessary and predictable ways. This claim, which for hundreds of years was seen as a fundamental precept of science, is a claim that has been rather thoroughly debunked by the cognitive revolution in the sciences. Instead of being machines, scientists have found that mind and inner experience exist in such a wide range of animals that it is safer to assume any animal has inner experience and intention than to assume that they do not. This is a really important point: we animals, all of us, live in and have perceptions of a world out there, and we have experiences of it in here. That is very fundamental but not trivial. Rather it is something foundational about living that we all share, the experience in here of a world out there. We all work continuously to harmonize the two. When we succeed, we are mentally healthy, and when we fail, we are mentally ill, humans and other animals too.

In the case of our lusty lions the implication of our shared circuitry for lust is this: the way lust feels for me is the way lust feels for most animals. Lust is compelling (to say the least) and it is pleasurable. There is a great range of behavior that goes with this basic lust urge in humans, and in other animals as well. Some animals, such as humans, dolphins, and orangutans, appear to force themselves sometimes on unwilling partners. Others—like swans and prairie voles, wolves and beavers and civets—mate for life in an affectionate partnership that forms the shape of their entire adult lives. And many animals are somewhat opportunistic in their expression of sexual behavior: they take what they can get, when and where and how they can get it.

Whatever the shape of the sexual behavior and its social bonds and meanings, it's clear that lust is a powerful force in almost every animal's life. Yes, we're used to talking about animals in terms of their "mating behaviors." It may be a little harder to own up to the fact that we too are driven by this thing called lust, that it's not just about reproduction, or "mating," for us or for them. The lions I saw were on a date, they were arranging a liaison. And lion liaisons typically last for hours if not days. They don't seem to have a notion of foreplay that leads to copulation, the point of it all, but rather an extended and affectionate mating session. It's been easy for us to project that kind of *animal* quality of sexuality, the idea that it is about physical conquest and physical release, on to the other animals and claim the turf of *relationship* for our kind. But these lions were in a relationship that was part of a larger set of relationships, and they slipped off to have a dance of intimacy together, a dance that is affectionate and lasts a while and that sometimes creates a lasting intimate bond of care.

The bond of care operates from the oxytocin system, but it is stimulated by any affectionate and trusting encounter, even by the memory or image of such an encounter. Not surprisingly such encounters are often in the context of lust and mating. As with lust, parental care is not the merely mechanical force of evolution once depicted by science.

When April the giraffe at Animal Adventure Park in Harpursville, New York, gave birth on live videocam, one of the zookeepers was quoted as saying that the two giraffes who kept watch over the enclosure wall were certainly females because males were basically only interested in sex and food.[9] I balked when I read that because I remember vividly watching as a wild giraffe was born in Kapama, and seeing how quickly it could stand with a little prodding from its mother, and how tired was its mother, and how its dad came and walked the little giraffe slowly within the cage of his legs, supported and protected, even as its umbilical cord dragged a bit on the ground. This went on for a while, and I have photo documentation that defies the zookeeper's pronouncement about how it is with male giraffes.

I'm willing to bet that, if you stop to think about how much of your own life energy and that of people you know has been spent either pursuing sexual experiences or relationships or attempting to control the urge to pursue them, or controlling what someone else is doing with his or her lust, you will likely agree that this single urge allows for a big window on understanding the other animals. For they, too, are compelled by it in equally complex ways. For them and for us, lust is a primary motivator. For us and for them, it's about mating *and* relating. Listen to Jaak Panksepp and Lucy Biven for a moment: "In all mammals, sexuality is promoted by provocative skin contact prior to direct sexual stimulation. Often, the prelude to satisfying sexuality consists of abundant playful courting activities, along with somatosensory stimulation that typically culminates in genital stimulation."[10] Recognize this? I hope so.

Can we really understand anything about these lions across the vast differences of our lives, us working in cubicles in towering urban centers and riding in cars of underground trains, them walking dusty paths through the bush? Well, yes, we can. Think about this: it stands to reason that if you've been an animal engaged in courtship, you can genuinely understand something about another animal engaged in courtship. And you may well also recognize—and feel in your own body—the utter tenderness of the care system, driven by oxytocin and generated by the "somatosensory stimulation" of sexual intimacy. Yes, animals share sexuality as a fundamental experience of living, including the developmental challenges brought on by sexual maturity, something also called "adulthood," something that must be created by each animal out of the chaotic chemistry of lust. The courting lions that walked off into the bush promised pleasure, and care, and new life because they had succeeded in becoming adult lions.

In Buffalo Camp we had deepened into the knowledge that you can't save the animals, difficult as that knowledge is, without caring about everything. Of course, we may have a favorite totem animal, one who is more likely to be big, as we like to imagine ourselves, but every big animal requires a network of millions of smaller beings for its existence. This was the centering point of the classic scientific ecology book by Paul Colinvaux, *Why Big Fierce Animals Are Rare*, that the big animals we find so exciting, the ones who inspire affinity and awe in us, depend on the millions of little ones. "So why are there fewer big, fierce animals than herbivores? Well, it has to do with the amount of energy it takes to be a large predator, but ultimately depends on the soil and its nutrients, climate, prey availability, human influences on wildlife habitat, genetics, disturbances, and so on."[11] The weather, the soil, the rocks, the little flowers, the scorpions, the moonlit hippos and the three-toed rhino friends, the humans in their jeeps, and the marula trees: they all (but for the fossil-fueled jeeps) nourish each other, they all depend on one another.

Just as the elements of the environment can't all be noticed or comprehended at once, the memories come like flash cards, the image of the mongoose, the civet, Liezel and her bug nets, Edward and his chameleon, the snorting warthogs with the cutest butts on the planet, Lente Roode and her long-gone cheetah, Sebeka. Every single animal wants for food and sex and temperature control, and exhibits many behaviors related to those wants. And every single animal wants for play and entertainment, for care and connection, and every act of care matters. Somewhere out there in the bush, a stone birdbath sits at the base of tree, placed there by human hands, waiting for the rain, in anticipation of the delight of little birds.

CHAPTER 4

It Starts with One

When I made arrangements for my second visit to Kapama, I did not know that the triennial CITES meeting (the Convention on International Trade in Endangered Species of Wild Fauna and Flora) was taking place in Johannesburg at roughly the same time. The manager of Kapama's elephant rescue program, Adine Roode, was away representing Kapama at the convention, addressing along with the international community the devastating problem of poaching of endangered species in Africa, and I would miss the opportunity to talk with her. But I did know when I booked an elephant-back safari for Diane and me that it would be one of the last of these safaris at Kapama's Camp Jabulani.

I had been reading about abuses in elephant-ride tourism and I was not surprised to learn that credible operators were discontinuing the practice. Though I understood well the reasons for this change, and though I would have made that same decision myself had I been responsible for an elephant-rescue program, I was still sad because I can pinpoint with precision the single best hour of my life: six o'clock on a February morning in 2008, just after I had arrived at Camp Jabulani for the first time. It's simply hard to imagine that the most wonderful thing one has ever experienced will no longer be on the menu.

On that sparkling summer morning back in 2008, I met Jabulani, the young bull elephant after whom the camp is named, and Paul Coetzee, the elephant master in charge of the morning's events and of the elephants as a group. The two of them, Paul and Jabulani, seemed to epitomize the possibilities for communion between humans and other animals. Indeed, it seemed that the humans and the elephants were simply nature as it could be, perhaps as it once was, a Garden of Eden. Although it seems odd to be able to select a particular hour or day from a long life as the best, it does well up inside me that way because in that hour I felt the tingling of the life within me embedded within a kind of raucous symphony of life around me, all swaying together to the rhythms of my own heart and breath and the sway of an elephant's back.

I would never have guessed that Paul, the director of this dawn event, was in his seventies at the time I met him. I took him for twenty years younger, slightly bowlegged in his crisp safari uniform, a bit grizzled in his attitude with us—but not with the elephants. African elephants are willful animals but they are also cooperative when it is warranted, even if it is sometimes cooperation delivered with creative twists. A human has to take time to figure out how to listen to an elephant, how to establish mutual respect, how to elicit cooperation when maybe the elephant doesn't feel like performing as the human requests. Does the elephant make an effort to understand the human? There is no scientifically certain answer to that question, but a consistency of cooperation without precise and predictable obedience would suggest that the elephant does make an effort to understand the human while, like the human, not surrendering his will as part of the deal.

An illustrative example of the sometimes entertaining cooperative non-"obedience" of African elephants was offered by Fishan, one of the bulls who came from Zimbabwe, who, when asked to shake his head, flapped his ears. It's not that he's stupid; it's that he's smart, and perhaps intentionally funny too. Paul had mastered the art of listening to the elephants, and his communication at every level was clear. He seemed a man without duplicity, a walking what-you-see-is-what-you-get. I think the elephants liked that. I know I liked it.

Paul rather brusquely directed us to line up on one side of the low stone wall so that the elephant keepers could bring the elephant family out from their night lodge to meet us, initially at a safe distance. Many of the keepers who worked with the elephants had traveled with them from Zimbabwe years earlier and had stayed on to care for them. As the elephants walked toward us, Paul told us how Jabulani had been found near to death, just four months old, stuck in a mudhole and abandoned by his family, of how Jabulani had been brought here for help, had been comforted to sleep every night by a sheep surrogate mom, how he had been tended by vets and by the Roode family. No one there had really known what to do for an isolated baby elephant when they accepted Jabulani, but they figured it out by trial and error.

Paul recounted Jabulani's near-miraculous recovery, which earned for both the baby elephant and the new camp the name Jabulani (Rejoice). They were still figuring it out, he said, but so far, so good. With his blond-haired arms and legs backlit by the rising sun and giving him a kind of general halo, Paul continued recounting Jabulani's early life, telling us that multiple attempts to return Jabulani to the wild had failed because the wild elephants wouldn't accept this baby who himself preferred the humans who had raised him and restored him to health. As Paul talked, the summer day was waking and the rest of the elephant family ambled along, one by one, in a long and very loose line, with child and even baby elephants scampering around, grabbing one another trunk-to-tail, tumbling in the dirt and kicking up dust. Birdsong filled the sky, warthogs snorted at the earth, and I heard monkeys beginning to chatter in distant trees. I felt completely alive in a living world of relations, all the way up and all the way down.

I was riveted by Paul's oration as he recounted the now-familiar story of Camp Jabulani's blended family of African elephants, a group of fourteen that had been tagged to become bush meat for soldiers in Zimbabwe but instead were discovered by Lente Roode in her search for a family for Jabulani. She bought the entire herd and moved them to Kapama, moving and employing their keepers, too, in the hopes that Jabulani might have the family he needed.

Keeping a herd of elephants is expensive, so the idea of the elephant-back safari seemed a great way for these animals to contribute to the costs of their rescue from the butcher, a win-win situation. Some of the elephants had already been trained in Zimbabwe to the task of carrying humans. So the elephant-back safari program and the luxuriously comfortable Camp Jabulani came into being, an experiment in rescuing elephants and in funding their care. Six of the elephants carried humans on two guided walks a day, at dawn and at sunset, accompanied by walking humans with rifles, by juvenile elephants, and by elephants just out for the fun of tagging along. The whole herd spent the rest of the day roaming the bush, splashing in the water, and coming home to their sleeping quarters at night. Paul told us that the elephants had a lot of freedom, that they came to the barn at night because they preferred it—though Tokwe, the matriarch, didn't prefer the barn one night, the night she eloped with one of the wild bulls of Kapama to produce the blended family's first new baby, a female that the humans named Limpopo.

How one thing leads to another! The way that life unfolds, decision by decision, requiring constant creative problem-solving, is something you feel in every tusky elephant breath at Camp Jabulani. Paul recounted the moment when Tokwe, now stepping into the early-morning line, had first met Jabulani, how the humans had stood by, holding their collective breath. To everyone's joyous relief, Tokwe sniffed over Jabulani and then placed the tip of her trunk into his mouth, a familial greeting. She had claimed him as her own. It had worked! Jabulani, at five years of age, finally had an elephant family, most of whom were not related to each other genetically.

In the wild, elephants live in family groups, led by a matriarch whose sisters and daughters and granddaughters form a family group. Interestingly, the matriarch in a wild herd is the oldest elephant, not the smartest, nor the sweetest, nor the wisest, not the most courageous. Apparently, elephant DNA has selected for experience; elephants want a knowledgeable and experienced leader. Tokwe was eleven years old when she was identified as the matriarch of this blended family (female African elephants live an average of fifty-six years in the wild, though they only average seventeen years of life in zoos).

In the wild, males detach from the group at adolescence, initially hanging out with bachelor buddies in what some researchers call a posse, but often living a solitary daily life outside of affectionate family visits and reunions.[1] Jabulani's new blended family had formed uncoerced bonds in an unusual formation for elephants: the males stayed with the larger family group, though they could choose not to do so, and one adult female, Setombe, seemed simply to prefer the company of the bulls to the usual sisterhood of elephant society. Adopted children and natural children grew up together in the care of the extended family. As the keepers in their crisp uniforms got the elephants into a line, Paul introduced us to Jabulani and Tokwe, adoptive son and adoptive mother, in the flesh.

Six of the elephants were saddled, while other adults prepared to walk without saddles and a handful of youngsters tagged along because it was a family outing. The saddles were fixed with leather straps and the keepers wore leather bags across their chests, bags filled with treats, used regularly to entice the pachyderms. Paul's talk moved to the topic of elephant anatomy. Some of the juvenile humans had become restless, and they had grown progressively louder, singing songs from *The Lion King* and raucously taking selfies. Paul shot them a warning glance as I moved closer to him in order to hear. He held Tokwe's great ear out from her body to show us its magnificent size, and he showed us the big veins in her ears, used as fans for

cooling her body. After offering Tokwe a treat to thank her for her cooperation, he moved to Jabulani, offered him a treat and then lifted and showed us
his deeply padded and sensitive foot, used of course for walking but also for
hearing. This is the famous foot that contains cavities for receiving seismic
waves that are neurally connected to the elephant's inner ear.[2] By way of this
specialized anatomy, elephants can hear for distances of up to twenty miles,
accounting for the mystery of how elephants seem to know when and where
to meet for a family reunion that has been called over distances of many
miles.

Tossing another treat into Jabulani's expectant mouth, Paul pointed out
the elephant's large flat teeth, used for grinding all manner of plant material
in the African bush. Jabulani nudged Paul with his trunk or took a little step
back away from the action, as if to say that if it wasn't a party, he wasn't staying. He wanted treats, that was the deal. Paul carried on, showing us the two
nimble digits at the tip of Jabulani's all-purpose trunk, the digits being a very
refined kind of fingers, used for small jobs and for grasping, the pachyderm
equivalent to opposable thumbs. The trunk itself can easily lift more than
seven hundred pounds while simultaneously having the capacity to smell
four times what a bloodhound can smell. It is in fact elephants who win the
prize as having the most acute capacity to smell.[3] Finally, Paul noted that
the elephant's trunk contains forty thousand muscles, compared to the 639
muscles in the entire human body. And then he was done talking, leaving me
longing to possess a trunk like one of these, if not quite willing to have the
anatomical structure necessary to support it.

The bored teenagers had drifted away a bit, chatting and laughing, so
I was now at the front of the line. Paul motioned for me to come out and
meet Jabulani close up. I asked him if I could touch the elephant. Paul briefly
looked me up and down, a big and hardy fifty-year-old woman in bush
clothes, vibrating with desire to know the elephants, then said quietly, "The
elephants are very sensitive; they know who likes them and who respects
them. You can touch them all you like." I have always appreciated that affirmation, of me and of elephants. I touched—tentatively at first—Jabulani's
ear, and then his leg, his foot pad. And then I hugged him with all my grin-
ning might. How do you hug an elephant? I chose his tree trunk of a leg, reminding myself of the proverbial blind men describing the elephant according to which part of its body each had touched.

Being forced to surrender my personal date with Jabulani to the growing line of other humans, I took the opportunity to photograph the elephant
youngsters at play. They played as children do, left to their own devices: they

pushed and shoved, ran and gave chase, they pulled tails and trunks, until someone got mad and someone cried. Then one of the mother elephants
swung her trunk around to lift a feisty juvenile off a squished and crying
baby. I noticed that the baby elephant's trunk was shorter proportionate to
her body than the trunks of the other elephants, nature's plan to keep her
from tripping over it before she mastered the art of using it. The liberated
baby ran to her mother's breasts, two round breasts between her forelegs,
reminiscent of a human mother, and suckled there, demonstrating her baby
body's knowledge of the soothing oxytocin flow. The other elephant moms
and aunties pulled the pile of kids apart, laying their trunks over them to
still them and to keep them in check.

The play of the little elephants that I photographed that morning was a
perfect illustration of the brain circuitry, neurotransmitters, and behavioral
patterning for play that all mammals share, one of the seven biological systems described in the research of Jaak Panksepp.[4] Kids play, all mammalian kids play. Left to their own devices, kids don't sit still in orderly rows of
chairs, at least not until they have roughed and tumbled, to use Panksepp's
descriptive language. Panksepp and Biven report that, across species, young
mammals typically engage in sessions of play that last for about twenty minutes, chasing and squealing, pushing, falling, laughing until someone gets
hurt and the play stops and the children separate, often returning to their
parents.[5]

So much gets worked out as young mammals learn in play the roles that
will match their temperaments and abilities throughout their lives. Contrary
to the popular myth that male children engage in rough-and-tumble play
and females choose caregiving play, male and female mammals engage in the
rough-and-tumble play with equal fervor. The alphas dominate, the runts get
hurt, the clowns joke, the mediators smooth over the rough spots. But the alphas learn that there is such a thing as too much, that they must temper their
robust natures with kindness, and the runts learn how to form protective al-
liances. Every individual animal learns in play his or her particular talents
and vulnerabilities in managing life in a social world. And if getting hurt is
play's price to be paid, laughter is play's reward. In the laboratory, Panksepp
found that young mammals who were given the option of social play took it.
In fact, he came to enjoy tickling rats who enjoyed being tickled.[6] An admit-
tedly cumbersome but useful way to understand the depths of the idea that
play is fun is to say that play has been selected by evolutionary forces to be
experienced as rewarding. Play, in other words, is an expression of biological
values: It feels good to play. Nature wants us to play.[7]

Young animals deprived of play in the scientific lab setting behave as if they have attention-deficit/hyperactivity disorder (ADHD). But when allowed to play, and the play period is over, the young animal is able to focus her attention. Animals consistently deprived of play as youngsters continue to behave as if they have ADHD and continue to crave play even as adults. Human children denied rough-and-tumble play, forced to sit in quiet rows or handed a passively entertaining smart device to keep them quiet, might also grow up to be adults who can't stop seeking play. The research on other kinds of mammals predicts that these adult humans might also be easily distracted and thus challenged in conforming to requirements of adult social living. Panksepp and Biven hypothesize that lack of rough-and-tumble play, the kindergarten built into our biology, is responsible for much of the epidemic of ADHD in humans.[8]

What is it like to be an elephant? I may not be able to imagine being big enough to pull a tree up by the roots or hear a message sent in seismic language from ten miles away by listening with the pads of my feet, but I know what it's like to laugh and play. I would likely fall on my face just trying to imagine a single appendage both powerful and nimble anchored to this face of mine, an appendage that allows me to smell more sensitively than a bloodhound, to shower, to caress my loved ones, and to perform precise manipulations of small objects. I may not begin to grasp the baby elephant's developmental task when that trunk that is small and flaccid at birth begins to grow and then requires the mastery of coordinating forty thousand muscles! But I have shared with that baby the general tasks of development as an intensely social animal, and I share his brain circuitry and neurotransmitters, and the social organization and behavior patterns that make his play and laughter fun.

People often report that one of their pets or even a wild animal is making a joke, and other people of course accuse them of anthropomorphism. Rooted in the principle of evolutionary continuity, I think it likely that animals do make jokes of various kinds because play is one of the deepest impulses of mammalian bodies and brains. It's also likely that anthropomorphism, the general tendency to project our humanity into nonhuman situations, colors our perceptions of animals in many instances. Scientific psychology has demonstrated that we project our own experiences of life into our interpretations of the world with regularity. Marc Bekoff, career animal researcher and animal activist, prolific writer and winner of the Animal Behavior Society's exemplar award, carefully constructs the case for understanding other animals in the same human ways that we understand anything: "The way human beings describe and explain the behavior of other animals is limited by the language they use to talk about things in general. By engaging in anthropomorphism—using human terms to explain animals' emotions or feelings—humans make other animals' worlds accessible to themselves. . . . Thus, I maintain that we can be *biocentrically anthropomorphic* and do rigorous science."[9] The same anthropomorphism that can mislead us can also truly help us to appreciate the experiences of other animals because, to the extent that we actually do share life experiences, that act of imagination allows for accurate empathy. In the words of Panksepp and Biven, "The human-animal bond, easily formed with companion animals, is strongly based on the fact that we share evolutionarily related social-emotional systems."[10]

The role of affect and imaginative empathy in communication is equally important for conversations between humans: words alone don't communicate much. Recall the frustration of listening to the robo-voice on an automated call, or of misinterpreting the tone of a text message, rather infamous examples of how words alone fail to communicate. But words conveyed with intonation and gesture on the part of the sender, combined with the receiver's capacity to imagine what is intended by that other, the speaker, can create approximate understanding between two internal worlds. It is the two sets of skills combined, expressive skills and receptive skills, that create understanding between two beings. The trick to understanding the world of nonhuman animals more accurately is to distinguish between fanciful projections of our humanity onto other animals and accurate empathy for their experience of living. As with many things, science helps by giving us some factual information on which to exercise our imaginations.

One way to get across the species divide is to play, to trust that what looks like play in other kinds of animals probably is play, and to grasp that what play feels like for us is roughly what it feels like for them. Except that we can't really grasp what it's like to grasp a tail with a trunk, so it is important to tune into differences as well as to trust the foundational similarities. One caveat: if you decide to play with other kinds of animals, it is probably wise to remind yourself that they bring other kinds of equipment to the moment in play in which someone gets hurt and someone gets mad. None of us can compete with the elephant's trunk. These elephant kids looked like they were having a blast until one of them threw his weight around a bit too much, and another one got scared, and everyone—with a little help from the grown-ups—went back to his or her own corner. Until next time. I could swear some of them were starting to titter already as I was called away.

The time had come to ride the elephants out into the bushveldt. I climbed up a platform ramp, received instruction from Paul on using the stirrup and keeping my weight balanced, then swung a leg over Jabulani's broad back. Isaac Mathole, one of the men who had accompanied the elephants from Zimbabwe, was already on Jabulani when I clambered aboard. I was startled by how rudely I landed, but Jabulani didn't flinch and Isaac just laughed. Paul adjusted my stirrups and then Jabulani walked off with me bouncing along on his back, absolutely gobsmacked (me, that is!), to wait while the other humans found their ways to other elephant backs.

Isaac and I chatted over his shoulder. He told me that he loved his life of working with the elephants, and asked me if we have any wild animals in California. Sure, of course, I said, mountain lions, bobcats, coyotes, foxes, rattlesnakes. As I described the animals of wild California, these seemed similar enough to animals he knew, but bears, what were they like, he asked. I held my neck stiffly and swung my neck and shoulders and head around as a bear does, all in one locked movement. I lifted back my big bear head, showing teeth as if to roar. He asked what sound they make. I told him that they roar. Roar? What kind of roar? So I roared like a bear, and the elephants immediately began to stamp and to trumpet in response. Oops. Isaac was laughing heartily as all the keepers moved quickly to soothe and to control the elephants. On Jabulani's back I learned not to put the bear and the elephant in the same room.

How can anything be at the same time as noisy as a sunrise in the African bush and as deeply calming as a sunrise in the African bush? That question, it seems to me, is at the center of the magic in those moments. Everywhere the world in all its forms was waking. I heard the shuffling hooves of zebras and giraffes thick on the grasses, the close munching of leaves. I heard the snorts of warthogs, the birdsong, the *work harder, work harder* of the Cape turtledove, the raucous distant baboon chorus, the chatter of monkeys up above. It was as though the first day of life on Earth had broken within me and in that timelessness was a deep stillness.

Before this ride on Jabulani's back, I had spent many hours rolling along dirt tracks through African sunrises, I had walked through the bush, accompanied by an armed tracker, but I had never experienced anything as resplendent as riding through the bush at daybreak on Jabulani's marvelous body . . . swaying to the rhythm of his feet as he kicked up dust, stopping when he wanted to munch the vegetation, feeling and adjusting to his motion and his awareness of the world. Early on, as Jabulani was in full stride, I accidentally dropped my two-liter water bottle. Jabulani deftly scooped it up, swinging it over his head and past Isaac, just to my hands. As I reached for it, it receded in the direction it had come from. I heard the crunch of the bottle as Jabulani decided he wanted to drink the water more than he wanted to hand it over. Or maybe the whole thing was his little joke? He tossed the emptied bottle into the air. Like others before me, I found Jabulani irresistible.

The guides with their rifles walked at the front and back of the line of elephants, and elephant babies frolicked in less orderly procession nearby. We crossed a streambed where a large gathering of baboons was having a morning bath and grooming session, and soon we came upon a still water hole. With the morning sun bouncing light blindingly from its surface, our eyes were at first unable to see the telltale bumps of an alligator that, once pointed out to us, we recognized. Immediately my heart and lungs shifted into high gear. With the sun now a little higher in the sky, the day warmed quickly and too soon we returned from what was for the elephants the morning shift. I dismounted amazed, changed deep within me by that one hour.

While I was mesmerized by the elephant's swaying motion through the landscape of elephants, a human (or several humans) of a more practically minded sort had been making a video of the whole production. We could of course buy one, personalized to include footage of us mounting and dismounting and a bit of each of us riding along. We were offered snacks and coffee while the video production happened, and then we had a viewing on a big screen. How could I resist images of myself bobbing along in ecstasy to a soundtrack of Toto's 1980s rock classic blessing the rains down in Africa! I've long since lost that video, I regret to say, though of course the technology is obsolete anyway. Yes, there's potential for cynicism in this tale of the predictable appeal of an old-fashioned selfie, but there is no cynicism that can diminish the sheer joy of riding on Jabulani's back at daybreak. It is an experience I wish that every person could have.

I returned in the evening to watch the baby elephants playing between the sunset ride and bedtime. I caught sight of Isaac and he asked me, from about twenty feet away and with mischief in his eyes, how the bear sounds.

On my second visit to Camp Jabulani I wondered whether having recently read several articles about abuses of elephants in tourism industries around the world would give me a different perspective on the experience. Honestly, though, I was hungry for a taste of the magic, an affirmation that it had been real. I was eager to see if Jabulani would even vaguely remember me, and I was eager to share it all with Diane, who had heard with probably boring repetition over the years my stories about my favorite experience ever. Jabulani did not remember me, or if he did, I had not much mattered to him. But he was still Jabulani, playful, comfortable with humans, leader of the walk. On this second visit to Camp Jabulani, I got to ride Jabulani a second time, this time with Diane and under the direction of Tigere, the new director of the elephant programs. Paul had retired in my absence, and then Paul had died. When I learned these things, I was saddened by the loss of the confident and agile—and seemingly youthful—man who had appeared to prefer most elephants to most people.

Tigere told the story of Jabulani's rescue and rehabilitation this time, and much of it was familiar to me. But some things at Camp Jabulani had changed, not the least of which was the already familiar news that the elephant-back safaris were soon to be discontinued because such safaris had become a lucrative industry and, as a consequence of financial incentives, wild elephants were being captured and abused for human entertainment. Adine Roode and the Jabulani team agreed that they needed to send the right message, that elephants should not be taken from the wild for purposes of human entertainment, even though the elephants of Camp Jabulani had been rescued from ending up on human dinner plates. It seemed that I was not alone in feeling an ache, a deep nostalgia already for the rides that would soon cease to be. After we had dismounted I asked a group of the elephant handlers if they thought the elephants would be happier without the daily ride duties. One of them said no, he thought that Jabulani especially would miss the activity and the procession of human beings that had been his life. I had expected them to say that, yes, the elephants would probably be happier just being elephants. But several people seemed to feel it was a sad thing for the elephants, a kind of loss of identity and structure, though all understood the need to support clear international standards.

Another change that had taken place at Camp Jabulani was the decision made that the elephants could no longer freely make their own babies, neither among this family nor by mating with the wild elephants of Kapama, as Tokwe had done to produce Limpopo. Sebakwe, the dominant male elephant, had produced five of the six babies born in Camp Jabulani, and he might have happily continued making little elephants, but the elephants are now on a chemical birth-control program. I was taken aback by this decision because I had been enamored of the adventures of Tokwe and of the blended family seeking new liaisons in the great outdoors. It seemed there was something more wild, more natural, in the elephants being allowed to breed as they saw fit. But the limited range for two large groups of elephants on Kapama would result in conflict and danger for all of the elephants. Their numbers had to be controlled to protect both groups from being harmed by competition and conflict created by bringing rescued outsiders into the territory of the naturally occurring Kapama herd.

There had been one more change in policy: Camp Jabulani had become a rescue and rehabilitation program for elephants injured or abandoned in the horrific poaching industry. Various animal welfare organizations began to ask Camp Jabulani to add injured, abandoned, and traumatized elephants to their successful blended family. Timisia, a little girl, was brought to Kapama in 2016 as a very thin ten-month-old, having been found wandering alone after being seen trying to suckle from her dying mother. Mopame was found wandering alone in nearby Kruger Park at just four weeks old. Kumbura was found in the same way, dazed and wandering alone, but she was severely traumatized as was evident from her level of continuous arousal, fear, and aggression. The elephants in the stable next to hers responded to her unrelenting cries of distress by soothing her. Both Lundi, the disciplinarian of the herd, and her son Mambo reached through the boards that formed the sleeping stalls to stroke little Kumbura with their trunks. She was soothed by them and has now become a member of the family. It took her much longer, an entire year, to consider keeping company with a human.

At Camp Jabulani, the choice has been made to give resources one by one to elephants in need, especially orphaned babies and children. Several have now been successfully introduced into the blended family. Add to these the six elephants born at Camp Jabulani, and consider also the annual cost of $600,000 to care for them, and the decision to use birth control seems imperative. But it also felt to me that some of the romance of the elephant family had been lost in these decisions, even though the stories of the orphaned babies were heartrending, and their healing care within the Jabulani family of elephants heartwarming. Surely the little ones deserved a family. The decision not to let the elephants breed naturally, seen in another light, was the decision to affirm the value of each abandoned baby creature, each baby left on its own as a result of human poaching. The message is important because it affirms that it's not acceptable to kill even one more African elephant because each one counts, not as an abstract part of some bigger number, but as an elephant-person in a species characterized by intense and complex social bonding.

For most of the people listening to Tigere along with Diane and me, the good news was that they got in before the elephant-back safaris had been shut down, and most of them were hearing the story of Jabulani and his family for the first time. For me, the changes charged the air in a manner similar to what I felt when Liezel had mumbled the words "global warming." I was now much more aware of the deep shadows from which this elephant family had emerged. African elephant populations have been decimated by poaching and this one horrible fact is the context that makes every single elephant count from a human perspective. Judging by the intensity of elephant social life, it is probably safe to assume that every single elephant always did count from an elephant perspective.

The World Wildlife Fund estimates that ten million elephants roamed the African continent in 1930. Today the count is just 415,000, including both the savannah and forest elephant populations. The *National Geographic* explains: "A booming Chinese middle class with an insatiable taste for ivory, crippling poverty in Africa, weak and corrupt law enforcement, and more ways than ever to kill an elephant have created a perfect storm."[11] Elephant poaching seems to have peaked in 2016, culminating in an 11 percent decline in elephant populations in one decade. This also means that a great many intelligent, emotional, and highly social animal children were abandoned and traumatized. Like us, they depend on the wisdom of their adults to learn and grow, but their adults have been murdered before their young eyes. My point is not to traumatize the reader, but the single most gut-wrenching image I have ever seen was one taken from a helicopter of a vast field of elephants with their faces hacked off by humans, left to bleed to death, with baby elephants wandering dazed among them. Keep in mind that these are animals known for their grief rituals, funerals, and memorials. Long after the death of particular animals, elephants return to caress the bones of the departed. Many African law enforcement agencies have now adopted a shoot-to-kill response to poaching in lieu of making arrests, and it appears to be having an effect. Poaching has begun to decline, especially in East Africa.

As I walked up the ramp this second time and climbed onto Jabulani behind Diane, my heart was heavier and far less innocent than it had been when I first met him. And I was once again shocked by the difference between the landscape squirming with life that I remembered and the current reality. It was a relatively quiet walk. Yes, the drought again. So much for my expectation of the Garden of Eden. I was seeing the death side of the equation of all this life, and I was disappointed. I had wanted Diane to have that incredible experience, and I had wanted to have that magic hour again, but magic hours don't work that way; they can't really be sought or bought. If anyone had ever told me I would feel disappointment with an elephant-back safari, I'd have said that, barring bad treatment of the elephants, it was not possible. I know well it is only the utter magic of the first dawn ride that made the second one seem something less.

That ride, for me, felt like the end of an era. But it also represented the beginning of a new era. One of the contemporary projects at Camp Jabulani is a scientific study to compare the Jabulani elephants with the wild Kapama herd by analyzing the amount of corticosteroids, the stress hormones, in dung samples. Guests can now participate as volunteer scientists, trained to collect dung samples for analysis in the lab. The results show no significant difference in stress between the wild and natural herd and the rescued, blended, and somewhat humanized herd, an extraordinary finding when you consider the historical trauma and social disruption experienced by the blended herd.

Another study underway at Camp Jabulani is the Elephants and Bees project, which has determined that elephants respond negatively to the presence of honeybees because they have sensitive areas around their eyes and ears where honeybee stings cause them misery. At the sound of swarming bees, elephants show fear and confusion and begin to back away: a finding that opens new avenues for reducing conflict between humans and elephants, using beehive fence lines to keep elephants out of crop fields.[12]

Camp Jabulani came into being, as many animal-rescue operations do, because one person found one abandoned child and decided to help him: Jabulani, around whom the camp was formed. My two visits with Jabulani, with almost a decade in between, under changed conditions of weather, and in the light of developments in scientific research and political organization, allowed me to see how much and how little we know of helping other animals. It showed me in this particular and complex setting what I had seen and heard over and over again from people doing the work of helping animals: that the work is a work in progress, a work of trial and error, of learning by experimentation and by conversation. And it is a work in which many fear making mistakes, being caught out doing the wrong thing. Who to help, how to help, how not to help, how much to help, when to let go, when to allow both suffering and hope to end for a given animal are difficult questions, made more difficult by the sense of being watched and reported on that so many animal rescuers told me about in near whispers. Even Kapama, with its licenses and credentials, its teams of top-notch veterinarians and researchers, has been subject to the kind of unsubstantiated social media criticism that is so easy to voice and so detrimental to the good work being done.

There is very little certainty in the work of animal rescue and sanctuary, and saving other kinds of animals is exhausting of human energy and resources. Programs must ever more creatively find ways to fund the work, often ways that involve saving and preserving animals while entertaining and educating humans, so that human feelings can be engaged and human behavior changed. And sometimes those programs have to let successful funding projects go, as happened with the elephant-back safaris at Camp Jabulani. I will be grateful for the rest of my life that I had the experience of riding Jabulani at daybreak and cavorting with that amazing "unnatural" blended elephant family. These elephants generously rescue each other, welcoming abandoned babies, soothing traumatized children, and demonstrating again that it's not just the humans doing the saving but the animals themselves.

The humans who make that possible are a special breed of human, indeed, and in them I have found kindred spirits. There are obstacles and barriers to their work on every side, and perhaps even in response to this, animal rescuers and protectors often exhibit a kind of stubborn hope. Everything may look dismal at times, but as long as they have an ounce of resource and there is an animal who needs it, the dots will connect. That dogged hopefulness becomes a cultivated virtue that I have seen repeatedly in a kind of physical choreography as rescue workers consider the odds, sigh, slump their shoulders, drop their heads ... then raise them up again, take a big breath, and turn their faces to the next thing they will do. And perhaps that is most of all what I have seen and learned in meeting rescuers around the world, the dogged hopefulness of helping the next one in line while we can, learning over and over the core lesson of being a relational being, as both humans and elephants are: that while the other is like me in significant ways, he is not me, or mine, but his own, to be understood on his own terms or not at all.

No mathematical comparison of elephant counts over decades can move us to act in the way that a single abandoned elephant child can, and that is something we share with them, that our feelings are what motivates our behavior and our feelings are awakened through our embodied senses. In the world of animal rescue, it tends to start with just one, and the hope tends to be that the singular "just one" is infectious, that others will catch and carry and spread the knowledge that doing just one thing to save just one animal is worthwhile.

Adine Roode, in her opening message in the Camp Jabulani brochure, speaks to the contagion of the "just one" phenomenon. Noting the perilous decline in numbers and the fact that (as of 2017) there were still more African elephants being killed for their ivory than there were being born, she says, "Few reading these statistics would fail to be moved. But most will invariably ask, 'What can I do? How do I, as one human, save an entire species?' In all the years of our experience with elephants, perhaps the most profound single takeaway we have to share is this . . . Just ONE makes a world of difference. One life saved. One decision made. One person committed. One lesson taught, and learned. One mindset changed. It always starts with just one."[13]

I can't help but wonder sometimes if it is perhaps wishful thinking that *just one* makes any real difference. And yet I know both in my own experience and in my knowledge of science that one can make a powerful difference to many, in ways that are not predictable at the time of decision-making. I can't help thinking that poet Ashish Ram has it right: "One song can spark a moment, / one whisper can wake the dream. / One tree can start a forest, / one bird can herald spring."[14] Let me share the argument I made over dinner in 2015 to Peter Singer, author of what is widely recognized as the founding philosophical treatise on animal rights, *Animal Liberation: A New Ethics for Our Treatment of Animals*, written in 1975.[15] Singer is an ethicist of the utilitarian lineage, a model of ethics that generally seeks the greatest good for the greatest number.[16] I argued that we can't solve problems with the same kind of thinking that created them. Therefore the goal of ethics, I said, should not just be what is best for the greatest number of beings: that kind of abstraction on a grand scale is what has created the problems of ecological destruction we now live with.[17] Another kind of thinking acknowledges that because all animal behavior is motivated by feelings, the singular can matter enormously.[18] Peter acknowledged the point about individuals and single actions as important loci for ethical concern, and, I think, left it to me to make that case. The singular sparrow, the lone little goat, the abandoned baby elephant, may be the image of the intimate knowledge within that becomes a fire in a heart—and then, because we are such social animals, that fire in one heart may become a fire in many hearts, and so the cultural tide shifts. It is the kind of process that brings about actual behavior change, especially as compared with abstract arguments. Most people are not moved to action by statistical reports; they are moved to action by feelings, and especially by relational feelings.

To write stories of people saving animals is an act of radical hope, as is the saving of each animal. It is the opposite of playing the fiddle while Rome burns. This saving of animals across species boundaries, caring about them in their own lives and their own ways of being in the world, is an expression of bonds with the natural world beyond human societies, bonds that must be made and strengthened if we are to have ecological hope for ourselves.

I often reflect on the fact that we simply do not know which of our actions will have what effects, so we must always act as if the next action were the one in our life that will have the greatest effect. That is an impossibly high bar, but it's worth considering. In the training settings where I once worked as a psychologist, I often heard from interns expressions of disappointment that their interventions didn't seem to work. To this concern I responded by telling them that they are one moment in the complex life of that other person, that they have the opportunity to provide an experience that is like a seed, and that they will rarely have the privilege of confirmation that their work worked; they will more often *not* see the seed germinate and grow. But that does not mean that it didn't matter, or that it didn't work. Looked at from another point of view, to turn away from the good that you can do because it may not change the world is a pathetic surrender to cynicism. Hope may not be easy, but at least it is dignified.

I have been to Camp Jabulani twice, once in innocence, once in loss of innocence. And now I want to go back again, this time to collect some elephant dung for analysis in the lab.

CHAPTER 5

Endangered

A visit by a human to the Hoedspruit Endangered Species Centre (HESC) is a thoroughly and carefully controlled experience. At first glance it seems as though the other animals are imprisoned behind their fences. But that is the lens we are accustomed to looking through. In fact, the animal enclosures are large, created as spaces in which the animals can be themselves, while yours is small, and it is you, the human, who is kept behind bars in the jeep rolling along between the double fences, seeing the animals at a distance meant to protect them, as you listen to talks about the endangered-animal programs. You are actually the one who is completely controlled here. You are the one who is not allowed to spontaneously express yourself but are held within narrow constraints. And with good reason, because you represent the danger after which the Endangered Species Centre is named.

This is what I was hearing, a message coded not in words but in the mechanics of the context, a message that penetrated more deeply than the verbal messages. It made me think of poet Robert Frost's line, "Before I built a wall, I'd ask to know what I was walling in or walling out."[1] The verbal messages were full of both welcome and information, delivered by a uniformed employee who spoke into a microphone as the green jeep, at maximum carrying capacity, rolled along that narrow paved road between the high chain-link fences that were double- and triple-gated. I quickly understood that HESC, though part of Kapama, is different, qualitatively different from Camp Jabulani, and also a necessarily different experience than that of exploring the Kapama conservation project with Aneen and Fossie and Diane. I found the difference to be oppressive at first. I was hard pressed to articulate the depths of it, but it comes down to this: the visitor's experience of HESC not only lacks the intimacy of Camp Jabulani, where the well-behaved might still hug an elephant, but it lacks, too, the sense of freedom we had felt rolling through the bush awake and aware, expectant. Here at HESC the goal seemed to be that nothing unforeseen should take place.

The employee continued to tell us about this place where we had found ourselves, just as we caught our first glimpses of the cheetahs for whom it was originally named. What is now the Hoedspruit Endangered Species Centre was once the Hoedspruit Cheetah Project, charged with scientific study of cheetahs, then a listed endangered species, and also with the conservation and breeding of cheetahs in the hopes of release into the wild. However, and not surprisingly, members of other species soon began to arrive. The mission was expanded to include animals other than cheetahs and the name was changed accordingly. HESC now houses vital rescue, conservation, and breeding projects for species whose continuity is endangered by humans: through human activities called development—the capture of wild ecosystems for human use—and through poaching, the illegal hunting of endangered species for financial profit. HESC now houses populations of African wild dogs, the second-most endangered carnivores in Africa, and cheetahs. Additionally, the orphaned children of poached mother rhinos come here for medical treatment and for healing, simply and profoundly recognized as beings worthy of effort and expense. HESC is a complex project by way of which some humans try to right the wrongs of other humans. It is not simple, and it is not easy; and it requires careful protection of the animals who live here, some on their way to the wild again, others dependent on human care for the duration of their lives.

With seven hundred cheetahs on-site, we learned, this single program houses an estimated 10 percent of all the cheetahs in the world today. HESC has successfully bred cheetahs and then released them into the wild to carefully selected conservation projects with adequate resources to support them, a significant accomplishment given the fact that cheetahs are not highly successful breeders on their own, even in the wild.[2] Cheetahs have low genetic variability because they tend to live and breed in small population pockets within large areas of range, and so the viability of offspring is low. A stunning 90 percent of cheetah cubs die within the first three months of life, about half to predators and the other half to the various con sequences of weakened immunity resulting from the lack of genetic variability. The cheetah breeding program at HESC controls for distinctly different genetic lineages in an attempt to produce more viable cubs.

Our guide pointed to the two kinds of cheetahs, the dark-spotted king variant and the lighter-spotted cheetah of the dominant strain. He explained that the cheetah babies born here are carefully prepared to live as wild adults. They eat the same meat in roughly the same quantities that wild cheetahs eat; they do not eat things that wild cheetahs don't eat. They are protected from contact with humans. In preparation for release the cheetahs are put through a process of adjustment or rewilding. HESC's website explains, "During this period the animals are transferred to large enclosed areas, which have a suitable prey base in a habitat of mixed open savanna and grassland, as cheetahs prefer open areas to hunt. Then the rations fed to the animals are reduced over time to entice them to hunt natural prey. The animals need to be closely monitored during this time to evaluate their suitability for possible reintroduction."[3]

Some of the cubs born at HESC have given birth to their own cubs in the wild, considered the measure of an effective captive-breeding program. Additionally, chances are that these particular cheetahs have a better rate of genetic diversity than cheetahs breeding in the wild and thus a better chance of living to maturity and breeding themselves. And so goes the slow building of human knowledge and competence in helping endangered populations to maintain their existence, a slow work in the context of the rate of extinctions, a kind of plodding carefully against the clock when our impulse is to race.

Beyond the problem of lack of genetic diversity in cheetah populations, the biggest challenge to the survival of the cheetah species is neither capture nor hunting but the disappearance of range due to what we humans call development. This is no doubt a factor partly responsible for the decline in genetic diversity because it forces the cheetahs to breed within small local pockets, generation after generation. The world in which the cheetah evolved, not long ago expansive and wild, allowed for long runs, fast runs, and for the successful hunts made possible by the skills in which their biology excels. In a world divided up by fenced boundaries and roads, not to mention significant artificial boundaries of various human legal jurisdictions, cheetahs do not fare well. While they enjoy protection in smaller settings like Kapama, they are designed for the larger setting of open spaces. The day comes when in order to survive as a wild species they must leave the safety of that place and move into the larger world for which they are made. For this reason, of course, it is urgently important that we create safe passages across physical and legal boundaries for animals whose lives take place in bigger geographical contexts than our own. International forums like CITES and the International Whaling Commission are especially important to species that require passage for their existence.

Our guide was now telling us that cheetahs are the fastest land animals on Earth. They go from zero to sixty miles per hour in three seconds, as fast a start as race cars, and normally achieve cruising speeds of sixty to seventy-five miles per hour.[4] They are introverts, males hanging out with a small number of adult siblings and living separately from females except during mating for the duration of their lives. Females are solitary except when mating or raising cubs, who are born the size of house kittens with a little furry mantle that disappears with maturity. Mothers stay with their cubs for eighteen months, actively teaching them to hunt and to avoid predators.

Adult cheetahs are thin with deep chests and huge hearts and lungs, enabling quick and efficient oxygenation for bursts of high-speed chase. Unlike other cats, they have small teeth to allow anatomical room for very large nasal passages that allow them to suck air into that deep chest. Their claws don't retract but, rather, act as cleats to give them traction. Amateur trackers often say that it can't possibly be a big-cat print if you see claw marks. Unless, of course, the cat who left the track happened to be a cheetah! The small teeth are a disadvantage in situations of conflict with their most frequent predators, lions and hyenas, but they can easily outrun these ... except as cubs. Cheetahs have unusually flexible spines and long tails that act as rudders, allowing them to change direction quickly in midair. What must it be like to be the fastest land animal, to have almost all your babies die in infancy, to wander across a jurisdiction line that is invisible and unsmellable and to find yourself suddenly captured, or, worse, shot and killed?

Can we know what it is like to be a cheetah? Well, we stay with our mothers for a long developmental period, and we share that bond of care as the foundation for social life. But can we comprehend the mother cheetah's experience of losing 90 percent of her babies before they reach maturity? Does she grieve over them? As far as I know, we don't know the answer to that question. Of course, we know what hunger is like, so yes, we can identify with the suffering of not being able to get food, and we can empathize with the pleasure of successfully getting food and eating it, though we prefer ours cooked. And yes, we can empathize with the experience of the discomfort of confinement, a state that is oppositional to that most basic shared animal instinct, the neurological seeking system that all mammals share, the system that biologically pushes each animal outward to engage the world. We can safely assume that confinement is an extremely uncomfortable state for most animals. But can we imagine feeling confined within the space of, say, one square mile? No, because that does not feel like confinement to us. A square mile is much too small a space for a cheetah, and it might well feel like imprisonment would to us. We cannot know in our bodies what it is to need several hundred square miles of physical space.

Animal bodies vary and animal sensory systems vary, and so the content of thinking minds shares the feature of describing the real world while varying in the aspects of that world that are salient to the particular kind of animal. Yes, we each perceive the world around us. We feel it, and we think about it, and we solve problems of living within it. We communicate about it, but our *way* of thinking is distinct. And this fact makes the similarity of our affective life, our feelings, all the more important. It is in the domain of feeling that we are most similar and in the domain of feeling that we can first and most easily bridge our species differences. Meanwhile, the basis for comparing one kind of mind to another is gone, because minds express the bodies they inhabit and the ecological niches they inhabit, and they solve the problems of those bodies in those niches. We just don't know what feels confining to animals evolved to cover enormous geographical distances with their bodies: cheetahs, migratory birds, whales. What does the whale think, living in the vast ocean, in a huge body that is virtually all ear, with sound waves amplified by blubber? I don't know, but I can guess that the whale thinks in sounds in a way that is more elaborated than my sound thoughts, and this would be true even if I were a professional musician or linguist. Whale thought, I propose, shares features of thought with mine in that it is an abstract representation of the world in which the whale lives, but it is distinct from mine because whale thought occurs in a whale brain and body and in a whale ecological niche.

It was long a habit among biological scientists to be misled by the assumption of human superiority, indeed by the very relevance of the concept. One example of this is in the domain of studying animal linguistic capacity. We spent decades trying to teach human languages to other kinds of animals because we assumed that only humans have true language and that some of "them" might learn from us language in a rudimentary form. But at last we have moved on to conducting observation and to developing, testing, and revising hypotheses about animal communication and language. The assumption that human capacities define the pinnacle or highest degree of any trait or behavior had limited our ability to observe things like elephant seismic communication, so different from human communication as to be incomparable. It is not a matter of degree, but of kind. They use sound waves in the ground to communicate; we do not. The assumption that human language is the best and most advanced language delayed our attempts—now well underway—to decipher animal languages.[5] In the context of ecosystem niches, the best communication system is the one that works for the purposes of a given species in a specific ecological niche. The same principle applies to a range of traits and behaviors.

The expression *differences of degree, not of kind* has been widely used over the past several decades to describe the general trend in scientific research on animal cognition and behavior, the most fundamental finding being that humans are not a different kind of thing from other mammals but members of the animal family of life, existing in evolutionary continuity with other animals to varying degrees. All animals fit this description: every species is continuous to some degree with every other species. But the very word *species* describes, by definition, differences of kind. And because both continuity and differentiation describe the essential features of the evolution of life on Earth, the expression *differences of degree, not of kind* is actually very problematic for biologists.

The expression originates in words penned by none other than Charles Darwin in his pioneering work *The Descent of Man*. Darwin held that "spiritual powers cannot be compared or classed by the naturalist: but he may endeavour to shew, as I have done, that the mental faculties of man and the lower animals do not differ in kind, although immensely in degree." Significantly, he added this caveat: "A difference in degree, however great, does not justify us in placing man in a distinct kingdom."[6] In this statement, Darwin was addressing the idea, promulgated by the philosopher Descartes and widely accepted as fundamental to the scientific worldview of his day, that humans alone have rational capacity.[7] Darwin was rejecting the notion that other animals are mindless programmed machines, and he was articulating that we are continuous with other animals.

But the usage of the shortened formulation *differences of degree, not of kind* is confusing. In reality, all animals are genetically continuous. Consequently we would expect to find the basic features of mind in all animals. What are these common basic features? That is a very good question, certainly open to debate, but I offer these: sensation, feeling, thought, and communication as fundamental aspects of mind in at least all mammals, if not all animals. And I further suggest that, in fact, the differences between my mind and my dog's mind are not a matter of degree but of kind: my dog's primary sensory modality is olfactory. I cannot really imagine what "smell thoughts" are like, but I can easily deduce that this is how my dog thinks. I read email, but my dog reads peemail—very attentively in her case, from the moment she emerges for her early-morning walk. Because she and I are both members of highly social species, we each find the receiving and giving of messages to others of our kind to be a powerfully compelling activity. There is continuity and there are real differences between us and across species as a whole. And so it is the second half of the expression *differences of degree, not of kind* that is problematic. There simply are differences of kind. I do not have integrated vision and hearing in my perceptual brain, as does a dolphin. The dolphin hears what she sees and sees what she hears. We humans do not. The bees see ultraviolet light and so have a much broader range of color vision, allowing them to see patterns on flowers that we cannot see.[8]

In short, I reject the expression *differences of degree, not of kind* because it tends to diminish the distinctiveness of speciation and the capacities of other animals. Evolutionary science posits two incontestable ideas: that all of life is related by way of common origins, and that life-forms diversify in order to better adapt to environmental niches. It follows that both the unity of life-forms and the diversity of life-forms are important dimensions for study and for critical consideration of the possibilities in human-animal relations. But the long-held assumption—or, rather, question—of human superiority has only served to obfuscate our understanding of other animals. I hold to the scientifically more accurate rendering, that all animals share fundamental structures and functions of body and behavior to varying degrees, but that there are indeed important distinctions by species, differences of kind worthy of our disciplined attention.

We can be curious about our human compulsion to ask the question about who is better, smarter, stronger, but let us also understand that our desire for superiority is irrelevant to the world and to other animals except to the extent that our related behavior impacts them. As Liezel Holmes suggested, you can try the personal experiment of walking up to the lion in the bush to ask him who is smarter, who is stronger, but you will probably reconsider that experiment as quickly as I did.

The central core of my life's work is to articulate the grounds for real relationships between humans and other animals. This implies the capacity to empathize accurately and to be able to communicate across species differences, but also the capacity to recognize that the other is not "just like me." There are similarities that provide for genuine cross-species empathy, and there are differences that open our minds creatively to the world in which we live. If we are to have real relationships with other animals, it is important to recognize these real differences and, instead of expecting these differences to distance us, to expect that they might enrich us. In some instances, the differences may be of degree, in others of kind, but the differences never erase the similarities, and the similarities do not erase the real differences. As evolutionary biologist Neil Shubin has observed, "There is order to what we share with the rest of the world. We have two ears, two eyes, one head, a pair of arms and a pair of legs. We do not have seven legs or two heads. Nor do we have wheels."[9] OK, maybe cheetahs have wheels, or even wings, and jet packs too.

The HESC cheetah research and breeding project is one example of a private nonprofit doing the relatively slower work of research to learn the answers about what works, what doesn't work, and why human interventions do or do not help other animals to thrive. The CITES meetings and policy documents address the bigger-picture issues of governments and nonprofits and private industry cooperating to preserve endangered species. If we want to provide breeding and rearing conditions for cheetahs, we need to understand not only their environmental but also their genetic challenges. These research and policy efforts represent the slow and sometimes grinding work that also allows for the wildlife veterinarians and anti-poaching police to respond effectively when emergency calls come in. If workers in the field are to transport injured and traumatized animals, they need to know how to do that; if they are to treat the complex wounds of poaching, they need clinical protocols; and of course they need a place to bring these animals, a place that will support the animals in recovery and often for the rest of their lives.

There is a fast side, an urgent side, things that happen quickly, that require immediate response and commitment, even when no one is really prepared for the problem. This was the case when the call came about little Phillipa, a two-year-old southern white rhinoceros, found next to her murdered mother who was pregnant. Rhino females reach sexual maturity at six to seven years old, so Phillipa was a young child when she was poached. She had possessed a young horn and the criminals who killed her mother also horrifically injured Phillipa, driving chain saws so deeply into her horn that her sinus cavities were exposed. They left her there to die with bleeding open wounds that had of course become infected by the time she was discovered. Phillipa, being a social animal with the same fundamental emotions and developmental processes that all mammals share, had just seen (and heard and smelled and felt) her mother murdered and had herself received brutal injuries, and Phillipa joined the ranks of—what do I say here?—persons who have post-

traumatic stress disorder, or PTSD. As both a psychologist and a philosopher, I use the word *person* for Phillipa, a nonhuman animal, advisedly, carefully. Though the majority of nonspecialists still disagree about calling other
animals persons, I have reached the conclusion that there is too much evidence supporting the application of the definition of personhood to nonhuman animals. I am not alone.[10]

At HESC, Phillipa has had several surgeries to heal her infected wounds,
to repair her damaged face. Wildlife veterinarians must invent treatments
as they are forced to deal immediately with unnatural wounds, so Phillipa
and others like her are in some sense experiments in treatment.[11] The treatments evolve in this way, to the point where Phillipa has even begun to grow
more horn material! And, of course, her successful treatment becomes a
treatment protocol for other rhinos with similar injuries. But the poachers also become more and more sophisticated as the financial rewards for
poaching become ever greater. With the joyous occurrence of Phillipa's horn
regrowth comes also the knowledge that this fact, the fact that her natural
face has healed, would endanger her again if her horn were allowed to grow
and she were free in the wild.

While her physical injuries have healed, the PTSD that Phillipa experiences is like the PTSD that human victims of horror experience: nightmares,
difficulty in relationships, fear and distrust, physiology that is locked in hy-
perattentive fear and aggressive responses to perceived threats, alternating
with numbing of awareness that allows for escape from the hyperaroused
state that so exhausts the biological resources of an animal's body.

We only saw Phillipa from a distance, from the jeep, through a carefully guarded series of fenced entrances and exits, with perimeter fences
equipped with powerful electrical currents sufficient to deter any would-be
poachers. The driver of our jeep stopped and pointed to Phillipa—deep inside the enclosed area—as she told us about her most recent surgery. I asked
if Phillipa still suffered from PTSD. Yes, she does, she has nightmares and
she keeps to herself. But Phillipa has finally made a friend in Ike, a young
male rhino rescued in Kruger National Park and brought to HESC by Saving
the Survivors, another nonprofit dedicated to helping the southern white
rhinos survive the onslaught. Ike had also been treated for horrendous
horn wounds, and other large wounds to his body, over a period of fourteen
months. Finally, it was decided that Ike's growth plate for new horn material
had to be removed, leaving him unable to defend himself in the wild, and
HESC agreed to accept him as a permanent resident.

Phillipa and Ike now have a friend in each other. The HESC website contains a touching typo, describing Phillipa and Ike in their togetherness:
"they are both very skittish, and weary of people."[12] Who knows how they
recognize another who has had the experiences that each of them have had?
We just can't imagine, nor do we yet have the science, to explain how one
rhino knows another who knows what it is like to survive their particular
horrors.[13] But these two recognize each other and, in that recognition, experience a moment of belonging, a small space of safety, a glimmer of hope
for healing their wounded souls. While some people continue to argue that
animals like Phillipa and Ike don't have minds and aren't persons, scientific
study and philosophical debate have demonstrated that they certainly do
have insides. As for souls, most of us easily understand referring to people
with PTSD as tormented souls, even without requiring a definition of soul.

Can I know what it is like to be a rhino, to weigh somewhere between
one and three and a half tons, to be nearly blind but to have a great sense
of smell, to have a tough hide and be virtually hairless, so that when I routinely run my two-ton body at the rate of thirty-five miles per hour on my
three-toed feet, there is nothing for the wind to ripple? Can I know what it's
like to have oxpecker birds as both security guards and groomers, routinely
sitting on my back or head, eating insects from my skin and making an ungodly commotion to warn me of danger, to have a small brain in a thick skull
and most of that brain given to the sense of smell? It's a way of life that seems
very strange to me. The closest thing I have to smell thought is the desire to
eat when I smell the garlic and onion in the oil, or perhaps even to have a
flash memory of my mother's kitchen when I smell a freshly baked cookie.
And perhaps most foreign to me is the fact that rhinos communicate with
their feces, scratching them into the ground and waving them into the air
with their tails, and also reading such writings as other rhinos produce with
their very refined noses. Our bodies and our environmental niches are very
different, and yet we and they know what it is to have a friend.

In fact, I would argue that I probably can empathize more genuinely with
a rhino's "soul" than with his body. Even though we share so much biology as
mammals, we also share a great deal of psychology. I have argued for a simple and natural (as contrasted with nonmaterial and "supernatural") definition of soul. In *The Archaeology of Mind*, Jaak Panksepp and Lucy Biven
make the case that "people and animals . . . feel that they somehow belong
to their world—that they are affectively embedded in the context of their
lives."[14] When a person or animal has PTSD, that feeling of belonging within

the flow of life is disrupted and the animal suffers anguish. To quote my own simple but profound definition of soul, "The feeling of belonging, of being responsive and responsible is important to living; it is not a mere illusion produced by mechanical gears but an interpenetrating organismic event. This is to say that from a scientific and objective point of view, life on earth is a genuinely shared experience and outside of that, it is nothing. I am inclined to think of soul as the experience of awakened awareness that knows that 'I belong to life itself, as do these others.'"[15] Humans and nonhuman animals who experience PTSD have their sense of belonging along with others in the world profoundly disrupted in what I would call a kind of deep soul sickness.

We don't know enough about rhino sensation, nor about how the rhino's combined senses create perception of their worlds, nor do we know enough about how rhinos interpret those perceptions, nor do we know enough about how rhinos communicate with one another about their internal worlds. What we do know is that it is more likely that they have perceptions, interpretations, and internal worlds about which they communicate with each other, and whereby they sense themselves as belonging, than it is that they are empty inside and we are the only ones with personhood. And while we may quibble over the definitions of persons and souls, Phillipa and Ike recognize each other in the safety of the HESC rhino enclosure, and each has a friend in the other, when both otherwise find relationships difficult if not impossible. Though they fear others, and especially humans, they do not fear each other. Together in their profoundly protected living space, Ike and Phillipa slowly heal the deeper wounds of the soul, recreating a sense of belonging with each other, as we do, by the very fact of sharing a history of those wounds.

Though differences between species are real, I contend, and the evidence supports, that humans and other animals, inhabiting different kinds of bodies, share a common psychological experience of soul. For example, the experience of basic fear, the anxious and deeply uncomfortable arousal in our bodies, the intense desire to escape, is something we share with all mammals. It's there for them when they are born, as it is with us; fear is biologically part of us, one of the seven primary affective systems described by neuroscientist Panksepp. Fear is just there in our brains when we are born, looking for a reason to be, ready to protect us from harm, to keep us alive. And fear is a very lively affect, however uncomfortable it may be as an experience.

All animals are genetically preset to experience fear. That little section of brain, the amygdala, functions from the minute we are born, whether we are a human or a reptile, to register the intensely uncomfortable sensations of fear and rage. To put this in the context of the cognitive development of an infant, you might contrast it with the human ability to acquire language, something that comes much later in human development, at about two years old, or even self-awareness, the ability to distinguish between oneself and one's environment, which comes online at about eighteen months for humans. Fear is there at birth, already a part of us, part of every animal, just kind of looking for an excuse to be . . . and then, of course, the world provides scary things and we begin to create our personal program of fear.

If programmable fear sounds like basically a bad idea to you, consider the whales, who seem to experience relatively little fear. It seems their fear went into a kind of dormancy because for millions of years they didn't have much to be afraid about. And here's the wretched news: that lack of fear is what makes them vulnerable to tiny soft little humans with ships and harpoons. This is how humans were able to hunt them almost to extinction. The whales wanted to play but were instead hunted; and whalers were terrified when those whales became angry, when they turned to smash the whaling ships to splinters.[16] And that rage response is the other capacity that is hardwired in the amygdala at birth. Rage is also a primal response to danger, but fear is useful because it protects us from danger in the first place, even if it protects us, too, from things that are not dangerous and provides us with intense discomfort that we can program into so many experiences in the process.

The way that primal fear is part of the biology of all mammals is beautifully illustrated by this description of the wolf pup in *White Fang*, written by Jack London and published in 1906: "Never, in his brief cave life, had he encountered anything of which to be afraid. Yet fear was in him. It had come down to him from a remote ancestry through a thousand lives. It was a heritage he had received directly . . . through all the generations of wolves that had gone before. Fear!—that legacy of the Wild which no animal may escape. . . . So the gray cub knew fear, though he knew not the stuff of which fear was made."[17] Humans perhaps elaborate the basic experience of fear more than other animals do, so that we experience more fear and more kinds of fear than do most animals. As Panksepp and Biven note, "We humans can learn to fear more things, past and future, than a little mouse can. . . . In multiple senses, we humans are the most fearful creatures on the face of the earth."[18] This has everything to do with the complexity of our programming. Other animals like the rhinos at HESC can experience PTSD, as we do, with intense or prolonged exposure to fearful stimuli, and they can become locked in horror. That's how PTSD works: it powerfully programs our capacity to feel the arousal of fear, and often the response of rage too. We all have primal fear that becomes programmed by frightening experiences, and then we form fearful and distrusting interpretations of the world.

This primal fear that we all share, though, is only one aspect of the complex condition that is PTSD. PTSD is the result of a terrifying experience that creates a kind of long-term traffic jam in the brain, a jam that does not allow for normal brain processes to flow, and so does not allow the animal who has experienced the terror to heal. That's why Phillipa has nightmares and why it took her a long time after being poached and rescued to find a friend. In more natural experiences of trauma, like those of a grazing animal chased and hunted by a predator, animals escape from danger, shake themselves vigorously, and avoid the places and circumstances associated with the threatening event.[19] Then they busy themselves with what is present in their worlds. Phillipa and Ike, like humans with PTSD, are very slow to recover the flow of mind that allows us to be present with our worlds, but instead tend, like us, to be locked by PTSD in the horrible past.

Back to my underlying question: Can we humans understand what it's like to be a terrified little rhino? Of course we can. To quote Darwin again, "Terror acts in the same manner on them as on us, causing the muscles to tremble, the heart to palpitate, the sphincters to be relaxed, and the hair to stand on end."[20] The wide-open eyes, dry mouth, tight gut, muscles drawing the body away from something dangerous that we feel when we are afraid is the same thing that they feel when they are in danger. When the leopard makes that indescribable sound—not a roar, but a loud whisper that says *I am here*—the bodies of prey animals go to that primal fear. The difference between us and them is that it's much more common for us to feel this intense fear in the face of things *imagined*, like being laughed at as we walk naked through town. But we can learn to use the adaptations the other animals use—to "not go there" with our thoughts, to shake our anxiety out physically, to orient ourselves to the present. Indeed, the mindfulness practice and embodied psychology that have become central to much of contemporary psychotherapy engage just these skills of presence to the current reality in which we physically exist. Yes, we can understand the frightened baby rhino, and even more, we can learn from the other animals. Maybe we can use what we learn from healthy fear response to help us help the ones like Phillipa who become like us, locked in fear of human making. It's complicated, but there's a common core of experience that is not complicated. We can understand fear in other animals, in Phillipa and Ike, and also in the ones who get to shake it off and get on with life in the present.

Meanwhile, the poaching of rhinos and elephants in Africa is creating more and more animal PTSD victims, especially the young who see their parents and families slaughtered and whose development is then derailed. This is why Liezel, the head ranger at River Lodge, had been unwilling to give even vague information about the number of Kapama rhinos. I was shocked by the sudden stop to her flow of enthusiasm when I asked about them. So I asked about poaching, and Liezel described some of the history of poaching and anti-poaching at Kapama, a kind of horrible evolution in which each side gains a temporary advantage, forcing the other side to innovative responses. Liezel explained to me that the guys pulling the trigger in the bush are paid maybe 3,000 rand (about $200) for a rhino horn that is worth 10,000 rand at the next level of exchange, and so on until it reaches a value of more than $300,000 on the Asian market.[21] It is the market for rhino horn, far away from the bushveldt where these animals are born, that threatens them with death, horrific injuries, disfigurement, PTSD.

At Kapama, significant resources are devoted to in-house anti-poaching police, and the Kapama police train other local police in anti-poaching efforts, while HESC provides community education to help local communities understand that their own economic well-being depends on the continued existence of the beautiful wild animals. Meanwhile, the poaching problem has become so extreme across Africa, with many poachers engaged in international markets, that Botswana has instituted a shoot-to-kill (poachers) policy, making it clear that if you come to Botswana to make your fortune in poaching, you should not expect to leave Botswana alive.[22] The two approaches are worlds apart: to educate the community at HESC, and to train animal protection police to shoot to kill poachers in Botswana and other African nations. And so the coevolution of poaching and anti-poaching continues.

The southern white rhino population had been rebuilt to an estimated twenty thousand animals, but in South Africa alone more than a thousand rhinos a year were poached for five consecutive years from 2013 to 2018.[23] Some argue that a complete and completely enforced ban on ivory trade is the best solution.[24] Others argue that private breeding of rhinos and sale of rhino horns on the open market would undermine the motivation for poaching, and still others suggest that the market should be flooded with current confiscated stocks of ivory in order to drive down the value of rhino horns, thus buying time to study the bigger issues. There are heated debates taking place among humans, while the rhinos must live behind layers of chain-link charged with electric current if they are to have the most fundamental safety, even a chance to reproduce.

Most of the people who work to end poaching agree that any study of the bigger issues would involve research on effective techniques for reducing demand for rhino horns in those parts of Asia where it is believed to offer a cure for disease. Techniques to reduce demand on the market side, like techniques to disrupt the sales side, tend toward both education and prosecution. But as has been found in the case of human prostitution, banning the purchase is generally much more effective than banning sales that depend on desperate low-ranking individuals on the supply side, willing to risk arrest and prosecution, even death, because they have so little to lose. Serious prosecution of those who purchase ivory and other animal body parts and sentencing with meaningful consequences would likely be more effective than prosecution of sales. This kind of policymaking on the demand side combined with policies to provide resources and alleviate desperation on the sales side might be the most effective combination to reduce and even eliminate poaching. One thing that is clear is that there is an urgent need for policies to reduce poaching, policies with enough consensus and buy-in to be effective.

In the meantime, the rhino population is being decimated. And little persons like Phillipa and Ike are suffering horrific injuries, both physical and psychological. Programs like HESC must make difficult choices about their use of resources. Should they try to help individual animals to heal? To help threatened populations to rebuild? To conserve and protect environments where rhinos naturally live? To educate communities about the value of the animals, and the plants, and the natural world that undergirds and supports all of them together? At Kapama, the answer is *all of the above*.

We saw Phillipa and Ike briefly, through all these multiple layers of protection, after seeing the cheetahs in their enclosures and the African wild dogs, then lazing in the shade of a big tree at noon, all from a distance, a safe distance. For them, from us.

Our last stop on the jeep ride was to the vulture space, as showstopping a spectacle as I have ever seen, a large fenced pit, about half the size of a football field. Around the pit are tall trees and perched in the trees are many vultures, too many to pluck at the bones at one time, instead taking turns at the hill of bones that seems to grow out of that pit. These are the bones of the animals fed to the wild dogs and the cheetahs, whose biology requires meat, and whose hope of return to the wild depends on their capacity to hunt prey. The terrorized rhinos, of course, eat grass.

The tour ended. We exited the jeep in a carefully controlled area that allowed us to leave through the visitor center with its historical posters, informational brochures, video displays, and gift shop. It was a big gift shop with sophisticated stocks of things to buy, fabrics and rugs, paintings and jewelry, books and field gear, to support the ongoing work of protecting endangered species. From us.

CHAPTER 6

Oxytocin Bay

Ever since my first acquaintance with the gray whales, I had longed to return to the lagoons of Baja California Sur in Mexico, to participate again in what must surely be one of the most enchanting of all interactions between humans and other animals—mother and baby whales seeking out groups of humans perched in small fishing boats, eager for the encounter. Size is no small part of the wonder. The boat is little, the humans are tiny in this context, and the whale is very, very big: an adult female Pacific gray whale weighs up to thirty tons and measures about forty-five feet in length. And she has traveled six thousand miles to give birth to the two-ton baby bumbling along beside her. She approaches the boat fearlessly, and as her infant whale gains experience, coordination, and skill, she will allow him to take the inside lane, between her and this boat filled with human toys.

The first time I visited these lagoons, I went to Magdalena Bay, one of the southernmost and less-populated lagoons. When I met my driver, Julia, at the tiny Loreto airport, I told her I'd come to see the whales. Her eyes took on a glow of devotion and she said to me, "Es un milagro! Las ballenas han venido a enseñarnos el perdón." (It's a miracle! The whales have come to teach us forgiveness.) She was referring to the horrible slaughter of almost the entire species in these lagoons in the nineteenth century—tragically reminiscent of what is happening in front of our eyes to the rhinos and elephants today—when whalers would harpoon an infant in order to draw in its mother, then slaughter the mother and haul her bleeding on board, leaving the infant to circle the ship and to die alone of its wound or of starvation, whichever came first.[1] I knew of this history, but I still wondered if Julia's response might be at least tinged with hype driven by the ecotourism industry that was our context at that moment. The whales have come to teach us forgiveness? Really? That was before I met the whales.

The whales of the late 1800s fought so hard against their hunters and captors that they were given the name "devilfish" for their capacity to destroy whaling boats and to kill the would-be hunters. And here is the extraordinary thing: many of the apparently friendly whales who began to court the humans in the early 1970s were alive in the slaughtering years, as evidenced by harpoon scars and easily deduced by their life expectancy of up to a hundred years. Yes, many of these same whales who bore the scars of migrations past were still alive for the restorative years of the new human protections now offered their kind. Many of the people living near the lagoons today have experienced the amazing curiosity and gentleness of the great gray animals, as mothers proudly show off their infants to human tourists. They marvel not just at the fact that these huge creatures seem to be interested in us, and to seek our friendship, but that they do so in the terrible historical context of this place where the whales seemingly now offer friendship, or at least harmless curiosity. Perhaps they have simply relaxed back into their giant pacific and unflappable natures.

The large-scale change in human-whale relations here in the lagoons has been attributed to human efforts to protect the whales, to institutions like the International Whaling Commission, to the status of World Heritage Site now afforded the birthing grounds, and to the thriving economy of ecotourism. All of these have no doubt made a difference, but the rebound of the gray whales must also be attributed to the resilience, complex intelligence, and profound sociality of their species. These whales had spindle neurons, a specialized long neuron that connects remote regions within the brain, fifteen million years before the emergence of the modern human species.[2] Spindle cells recently topped for an academic moment the ever-dwindling list, now abandoned by most biologists, of things thought to be uniquely human. In the case of the resilient gray whales, their own brains, behavioral capacities, social organization, and cognitive flexibility are reasons for their flourishing again, and part of the explanation of their motives to make relationships with humans. It comes down to this, that it's not only what humans have done to save the whales but what the whales themselves have done that has created this miracle of population recovery and also the concurrent miracle of interspecies—dare I say it?—love. Although it may seem an extraordinary claim, the scientific evidence supports it, and those who study the whales concur that it is hard to escape the conclusion that these complex, social, and intelligent creatures want good relations with humans, at least in this one area of their life space, the Baja lagoons.

Biologist Roger Payne, a career whale researcher, notes the relaxed confidence of whales and relates it to their size: "As the largest animal, including the biggest dinosaur, that has ever lived on earth, you could afford to be gentle, to view life without fear, to play in the dark . . . and to greet the world in peace. . . . It is this sense of tranquility—of life without urgency, power without aggression—that has won my heart to whales."[3] I can only concur, and I think it is safe to say that there is a consensus among those who have spent time with whales that it is indeed so.

Most scientists who study animal cognition and behavior have come to expect that other animals, to various degrees and in various ways, share features of mental life and behavior that humans experience, an assumption known as the principle of evolutionary continuity. Though this principle has been voiced since Darwin's day, it has taken us humans a while to accept the reality that other animals have inner lives too. Our pursuit of the philosophical idea that humans are above all the others and our scientific search for the source of that superiority are now considered defeated projects, based on false assumptions. After decades of research in behavioral science and brain anatomy and in the entire domain of emotion and motivation, the scientific consensus is that there is no such particular something that elevates us above all others in the animal family.

One of the more recent attempts to search for our supposed unique superiority occurred in the area of language, where many people still think it obvious that we must be superior to other animals. After all, they don't read and write books! Whales, however, do create songs in rhyming verse, as discovered by bioacoustics researcher Katy Payne.[4] Tecumseh Fitch, a leading researcher and theorist of linguistics, concluded one such search with these words: "These new data tell a cautionary tale: we must beware of considering any human trait unique without a thorough search among animals."[5] The new questions, based on a different set of corrected assumptions, are more like, "What are these creatures doing, and thinking, and feeling, and how, and why?" And, to bring it all back to our whales in the Baja lagoons, whale brains are the biggest brains on Earth and, some biologists argue, likely the best brains.[6] They aren't just big because whales are big, but they are big proportionate to body mass, and they also have the biggest gyrification index, a calculation that neuroanatomists use to measure the surface area of a brain as if it were unfolded.[7] As a result of all this new information, a new question about whales has arisen: What are they doing with those excellent brains?

I chose to return to the Baja lagoons during the month of March, three months after the first babies of the year were born. I knew that they would just be learning as toddlers the skills they would need for their first long migration—yes, that six-thousand-mile trip "back" to a home they could not yet imagine, a journey that would begin the following month. March is when the toddlers get into the inside lane regularly, when they begin to express their own interest in the humans. The early mornings are kindergarten in the lagoons, as the babies swim longer distances and encounter more challenging situations of tides and winds, in the company and under the direct tutelage of their mothers. By late morning, the lessons are over and it's playtime. I especially wanted to be there for playtime, when the mother whales bring their babies to visit the pangas filled with enthusiastic humans who want nothing more than to admire the babies, perhaps even to touch a "little" whale. And the toddlers, of course, are playful, sometimes actively resisting mom's attempt to call a time out. In this case, the panga captain is likely to support mamma whale by motoring away from the pair.

From the human side, this is a carefully monitored ecotourism operation, a thriving industry in Baja California where the whales come every winter to the shallow ocean inlets of Laguna San Ignacio, Ojo de Liebre, and Magdalena Bay. San Ignacio and Ojo de Liebre are within the boundaries of the El Vizcaíno Biosphere Reserve, designated a UNESCO World Heritage Site because of its vital importance to the thriving of the gray whale. Every winter the eastern North Pacific gray whales migrate from their primary feeding grounds in the Chukchi Sea, between Siberia and Alaska, to the Baja lagoons where they birth their babies.

Pangas, the small fishing boats that typically carry no more than twelve humans, are about one-third the length of a mother whale, and are restricted by number, by hours on the water, and by behavior of passengers. The captains are licensed, they know the rules, and their livelihoods depend on their keeping and enforcing the rules. In my experience, they also love and respect the whales. They do not pursue or crowd them. If there is something especially juicy going on and there are already two boats near to the whale making the juice, they let it be. The juicy thing might be a whale breaching, jumping clean out of the water to soar in an arc back into the sea. It might be a baby whale gently turning a boat by its prow, or maybe a spy hop, when an adult whale holds his head vertically above the water in order to see what's happening above the surface. It might be a whale mating event, typically one large female and two or three smaller males, a style of mating better understood in the whale context than the human: when you think about the logistics of needing to breathe in the water while mating, you begin to understand the need for support crew.

Whatever the cause of the human excitement, other captains must back off and wait for the next opportunity. In addition to not crowding the whales, there is no feeding of whales, no pursuit of whales, no swimming with whales, no harassment of any kind. Captains can go to where they expect to see whales and wait for them there. It's up to the whale to come closer to the panga or not, so when the whales come to visit, you know it's because they want to meet you. Pause to consider this, that the whales want to visit with the humans, as evidenced by the fact that they do it over and over, year after year, whale after whale, with no food involved. Their food is not in the Baja lagoons but six thousand miles away in the Chukchi Sea. That's right, they have virtually no food during the long migration. They visit us because they want to.

I learned during my first visit that the ocean sanctuary that makes these encounters possible—that has supported a new interspecies cultural event—was established in 1988 on the foundation of a commitment to protect whales from commercial hunting that began with a first international agreement back in 1937. At that time, in this very place, these vulnerable whales were being hunted almost to extinction. Their population had dropped to below two thousand individuals. The tide began to turn with the 1937 International Agreement for the Regulation of Whaling. When the International Whaling Commission was established in 1946, the Mexican government embraced its objectives and strategies. In 1973, the eastern North Pacific gray whale population came under the protection of the U.S. Endangered Species Act, which gave protection to whales migrating along the shores of the western United States. By 1994, the eastern North Pacific gray whale had rebounded from the brink of extinction, and it was removed from the U.S. endangered species list.

The eastern North Pacific gray whale is now considered one of the world's great conservation success stories, an icon of animal conservation, with a robust population now somewhere between eighteen and twenty-two thousand individuals, the high number representing what marine biologists think may be the current ecological carrying capacity. There are still threats to their thriving, with the effects of global warming on Arctic food supplies at the top of the list, followed by increased development of coastal lands and increased shipping and fishing in coastal waters. For eastern North Pacific gray whales, moving further from human habitation is not an option because they are shallow-water animals who navigate by the coastline. The whales have benefited from laws against their being hunted as well as from new protections afforded in their natal lagoons. Still, from a purely scientific point of view, the story of their recovery from near-extinction is remarkable, and it has become an icon of the possible. From the perspective of hanging over the side of a tiny boat, bobbing up and down on wind-whipped white-caps, putting your arm inside the mouth and stroking the baleen of a huge and more-than-willing whale is nothing short of amazing.

From the time of my earlier visit to Magdalena Bay, when I had learned the story and experienced for myself the thrill of groups of humans and whales playing together, I had wanted to meet Francisco "Pachico" Mayoral, the first fisherman to experience this new sense of connection in the 1970s. But the distance from Magdalena Bay to Laguna San Ignacio is just over two hundred miles as the crow flies, and a very long and winding road trip over isolated dirt roads for anyone who might find a willing driver. So, at my Magdalena Bay beach camp, on a moon-dark and star-bright night, to the sound of whales breathing over the quiet waters, I set my inner compass to return one day to San Ignacio to meet Pachico. I wanted to talk with him about his first impressions and about the ways in which he was received by his human community as ambassador to the whales in the days before the interspecies culture had developed.

Pachico died within that year, so the conversation was not to be. But a few years later, when I booked my trip to San Ignacio, I was surprised and delighted to learn that Pachico's son Ranulfo was to captain the panga, the eighteen-foot *Dolphin II*, on which I would encounter the whales again.

When I arrived, I looked at the line of windblown tents that would be home for the next few days. In my mind, San Ignacio was a legendary and therefore great place, so when I approached it in the hired van I was surprised to find that the Mayoral family village was in fact very basic and remote, consisting of a mere three houses approached only by a rutted dirt road across salt flats, with no grocery store or doctor or pharmacy, much less gas station or movie theater, within easy driving distance. Our whale camp had been erected for the tourist season just a few hundred yards past the family homes.

The setup looked similar to the camp I'd so happily experienced at Magdalena, completely if minimally functional, without running water or electricity. As I stepped out of the van, I looked down and found at my feet a piece of whale-shaped deep-blue sea glass. I put it in my pocket, taking it as a sure sign that my hopes for this visit would be fulfilled. I headed for the mess tent to receive instructions for living here and getting along with my fellow travelers, for cooperative relations with local law enforcement and with Mother Nature over the duration of our stay: how not to get burned, dehydrated, bitten, stung, drunk, or otherwise indisposed.

The mess tent was a long structure made up of several interlocking canvas tents, with a row of folding tables down the middle, both sides lined with folding chairs, a cooking stove and coolers and serving table at one end, and a path to the makeshift toilets at the other end—as if to model an alimentary canal, food at one end, compost at the other. The interior walls were decorated with illustrations of whales, whale migration, whale reproductive cycles, and comparative whale anatomy, along with a map of the lagoons and posters featuring local flora and fauna. From the inside support structures of the tent hung soft solar lanterns under which we would receive evening whale talks.

There I first met the others with whom I would share the next five days, beginning with four women from Minnesota, two of whom were regular adventure-travel enthusiasts, their less adventurous husbands left regularly and happily at home. Their equipment and the efficiency with which they used it were a clear testament to their experience and hardiness. These two served as both companions and support team for their more tentative adventuring friends, one of whom had brought with her an extensive makeup kit. A married couple, Ralph and Sarah, architects from the East Coast, had come seeking that deeper connection to life as part of her recovery from a cancer that had been almost fatal, but then not. She was washing up in wonder on the shores of life again and he was there to accompany and support her on her return. And then there was another solitary woman traveler, another Anne. Arriving from Köln, Germany, she was the only one who had come as far as the whales to be here, almost six thousand miles. She was in her seventies and loved red wine and cigarettes, seemingly second only to the animals around whom her life had revolved for a number of years. I later learned the extent of this, that she'd sold her home and was managing life in such a way that she hoped to spend all her money living her soulful and determined pursuit of close encounters with other species, hoping to get the timing just right before "throwing the spoon," as she referred to her own death. I told the group that I had returned to San Ignacio to conduct research for a second book on human-animal relations.

That first evening we met our panga captain, Ranulfo Mayoral, son of Pachico, and our crew, which included none other than Monica, a former dolphin trainer from Mexico City who had been working with the whales in Magdalena Bay on my first trip. It was she who had taught me to use a signature whistle to call to the whales, and it was my signature whistle that called Barbilla, my baby whale friend, to me, day after day in Magdalena Bay.[8] Monica and I remembered each other at once when we met here five years later in a different place, because we had shared the extraordinary experience of receiving a whale kiss. A whale kiss consists of the whale coming up out of the water, seemingly to make contact with a given person, head first and close, so as to be face to face, allowing for the human to plant a kiss on the whale's face. When Monica climbed into the panga that morning she had shared her dream from the night before, a dream in which she finally received a whale kiss. Indeed, she was the first to be kissed that morning, and then the little whale came to me for a kiss, perhaps the most memorable kiss of my life.

Ranulfo confirmed what I had heard and read about his father, that Pachico had come to believe that the whales can teach us about conflict resolution because they now engage the very beings who once slaughtered them. In the course of his surprising life among the whales, Pachico had moved from fearing them to saying that he would give his life to save a whale. Yes, Ranulfo related, this story of transformation, both in Pachico and in the lagoons, had a rare pivotal and defining moment. One morning in February 1972, when Pachico was fishing for bass with his partner in San Ignacio lagoon, a whale began to express interest in them. She swam around the panga, at first terrifying the fishermen, lifting her head on one side and then the other. At some point she swam under the boat to deliver what has become known as a whale hug, gently lifting the boat from the water and setting it back down. These behaviors now occur with enough frequency that there are names for them, "whale kiss" and "boat hugger," terms that imply there is a pattern of behavior between whales and humans that can only be called an expression of culture, an interspecies culture, something created in recent decades by humans and whales together. Typically we humans have referred to such cooperative relations with other animals as domestication, but if that word were to apply at all here, it might be said that the whales are domesticating the humans.

Back in the days when Pachico first noticed that the whale seemed to be courting his attention, the fishermen of the lagoons feared and routinely avoided the whales. But Pachico reasoned that, if the whale had wanted to kill him, it would have succeeded. He became more curious than afraid, and on that particular February day he reached out first one finger to touch the whale, then a whole hand; and finally, with his whole arm, he unstintingly caressed the whale. That day he learned not just to sit in the water with a whale but to reach back, and he began to spread the news that the whales were not to be feared. He said that the whales taught him how to relate to nature, and especially to animals, differently. "They were attacked by man, and yet they wanted to get closer to us."[9] There may have been other fisher-men and other whales, but we know about the one fisherman and the power of his connection with the one whale on that one day.

Pachico never ceased to be a simple fisherman, though he became a very famous simple fisherman, one who had a great love and who left a great leg-acy. Still, he lived in an earthen-floored home on the shore of an isolated and rugged lagoon, fishing and developing whale-watching as a local industry. In fact, he who said he would die to save a whale lived to save them. In addi-tion to unwittingly beginning the extraordinary interspecies event that now occurs every winter in the Baja lagoons, he helped to organize the shutdown of a proposed desalination plant on the delicate San Ignacio lagoon so that the whales might continue to come there to deliver their babies and to court the humans, so that the story could continue to unfold.[10] That first evening, over a beer in the orientation tent, Ranulfo promised that we would spend some one-on-one time talking about his experiences and perceptions of the whales and their interactions with humans.

We went off to choose our tents and I chose one nearest to the shore. I had paid to have a tent to myself, so I had to manage by myself when the whole thing blew over during a storm on that first night. I wasn't able to un-zip the sand-caked, rusted zipper, so I propped up a pole sufficient to pro-vide breathing space and to avoid direct body contact with the wet canvas. Then I got back into my cot and slept till dawn, when I emerged to see just how blown over was my home. I don't mind being without water and elec-tricity in the least. Rather, I relish opportunities to live closer to life ... un-til such opportunities involve serious loss of sleep. I long ago declared that sleep is my primary social responsibility, and it remains true decades later. I immediately began negotiations for a better shelter and was told that I might be able to move into the new hotel that Ranulfo's uncle was building a few hundred yards further down the peninsula. I was warned that the ho-tel was not exactly finished but that it did have a flush toilet and electricity, to which I replied that I only required that it not blow over. As I hauled my gear, daypack, camera, and life vest out to the lagoon for my first morning with the whales of San Ignacio, I begged to be booked into that hotel, sight unseen, in my absence.

Then I began to wade out and out through shallow water, surprisingly choppy for its lack of depth, over sand, rocks, shells, and old bones to where the *Dolphin II* and Ranulfo awaited us. Because the waters were so shallow and wind-whipped, we had to form a human chain to get the gear loaded onto the *Dolphin*. Ranulfo, in his yellow slicker coveralls and boots, hopped out to receive one armful after another of lunch supplies and safety gear.

The other Anne was unsteady on her feet, but refused help. She did not like being treated as a helpless old lady, even though a little help seemed called for. It took me a couple of days to stop offering. She later confessed that she'd been in a car wreck that was her own fault, that she'd hit a pole and had come very close to killing herself. She'd ruptured internal organs, broken twelve ribs, and had a spine that was now bolted together like a ladder. But one thing that became quickly clear was that she took big risks and also complete responsibility for herself. She, too, had come back from the brink with steely purpose, stark honesty, and a toughness betrayed only by her passion for and tenderness with other animals. Once on the boat, she told us that she was driven by the desire to touch again a mother whale. With tears in her eyes, she recounted to us that she had had some of those moments that are always called intimate by humans when we have them, of gazing into the eye and touching the body of a whale in a moment that feels eternal. She wanted to have that experience again, if she could, before leaving this planet. I understood the importance of that contact to her, I thought, and I told her that my hope was to look into the eye of an adult whale, as I had once looked into the eye of a baby whale, my little two-ton friend Barbilla. We two Annes recognized one another. Sarah was surprised that we had such conviction that the whales were persons to whom we could relate. She wanted to see what she would make of her own encounters, and she wanted to try out thinking of them as persons, a possibility she had never really considered.

Ranulfo steered the *Dolphin* out into the middle of the lagoon, turned west, and continued to move slowly until we reached sufficient depth to crank up the speed. Then we sailed out close to the mouth of the lagoon, where I stood in the back of the panga, exuberant with expectation, bobbing along in seemingly empty waters. Monica showed us how to splash the water with our hands from the side of the panga to solicit a curious response from any whale that might be lurking nearby. When we began to see whale "footprints," those round patterns on the surface of the water that tell you where a whale has recently been, she taught us to whistle for the whales. I stood and whistled, my old signature whistle coming back to me in this familiar setting. "Whale! Twelve o'clock," Ranulfo shouted and turned the boat in that direction. And so began a very active morning of watching whales from a distance, so many whales! Breaching and spy-hopping and zipping along in distant lines parallel to the shore. We saw so much activity that morning that we couldn't help but be satisfied, though no whales approached our panga. Perhaps, Monica pondered aloud, these whales were excited about beginning the return trip to their summer home.

Ranulfo pulled the *Dolphin* up to the shore of an uninhabited island and tied it to some rocks, while we hopped from the boat quickly and scattered out in all directions in search of a bit of privacy. That proved to be a tricky find on this sandbar with so little vegetation! But having taken care of one set of biological needs, we quickly moved on to the next, a lunch of guacamole and chips, fish tacos, orangeade, and water. I walked a long way in solitude along the shore, examining dolphin skeletons, some almost complete, others in parts, remembering that death is part of life, even as seabirds tore at any remaining scraps of meat or sinew on the bones as if to demonstrate where the continuity of life is located.

The waters were calm and we were calm when we returned to them to head back to camp. But our well-fed somnolence was soon disturbed. "A whale! Three o'clock, three o'clock," someone cried as, for the first time, a gigantic adult swam up alongside us about ten feet from the boat, her blowholes and mottled skin visible. It had taken me a while to learn the visuals of gray-whale anatomy, partly because they do not have fins on their backs, but instead a series of knuckles, visible vertebrae which, in addition to blowholes, tell you which side is up. Their nostrils sit undaintily on the tops of their heads, ridged exposures that allow you to navigate visually forward and down to the huge jaws that open into baleen-lined mouths, the baleen shorter and more bristly than I expected, looking and feeling like a cross between teeth and several layers of old tattered shower curtain. From the whale's mouth you look slightly up and back to find an eye (I have never seen two eyes at once, in spite of having looked at hundreds of whales!), and from eye to ear, the least easily located anatomical feature. And I cannot help but recite poet Homero Aridjis's lines to myself: "And the whales came out / to catch a glimpse of God / between the dancing furrows of the waters. / And God was seen through the eye of a whale."[11]

And here she was, blowholes and barnacles, with fluke and head beneath the water. Now that I really recalled what they look like, I could orient myself to them again. I had been unpleasantly jolted when I recently read this description of a gray whale: "Among the most ancient of all the whales, grays are also by far the homeliest, their gunmetal bulks encrusted with barnacles and lice and the crisscrossed scars of everything from orca attacks to the blades of boat propellers."[12] I was shocked to read this because I find them so very beautiful, but yes, the writer's description is accurate, I suppose—and he, having visited many species of whales, has a basis for comparison that I do not have.

We returned to camp and I used the afternoon siesta to move into my plywood hotel room, where I joined a company of maybe a hundred bees. I turned on the overhead light, charged my phone, used the flush toilet, stepped on a bee, stepped on another bee. And so violated one of the rules for cooperative relations with Mother Nature. I was barefoot. "Charlie bit me," I texted home, referring to the 2007 YouTube video of British baby Charlie biting his toddler brother, a family favorite that has become the literary reference for the variety of bites and stings that I report from the field. Having declined the offer by management of a can of poison insect spray, I spent the next hour systematically opening and closing the three windows in order to encourage the bees to leave. I also swooshed them out with a towel and carried them out in my hands, bees and me uninjured. It was only the stepping on one that brought an objection as serious as a sting. When I returned in the evening, I began to understand that the bees arrived in the morning as I departed, and they departed in the late afternoon as I returned. I began to get it, that I needn't work so hard, but I continued to wear my sandals to the shower. Not all of the bees had been hardy enough to make the afternoon flight, and those on the floor awaiting their deaths might yet sting.

We gathered in the evening for dinner and whale talk in the mess tent, and we had the first of many conversations about what it's like to be with these huge others. Outside the tent, I sat next to Anne sharing a glass of red wine while she smoked and told me about the accident. She also told me that her full name was Anneliese, but that she liked to use the simpler and shorter form when traveling. After dinner, Terry, owner of the tour company with twenty years of experience in Baja, talked about the whales, pointing to the charts and introducing us to some of the science, some of the lore, and some of the uncharted spaces in between the two. He reported that a Canadian visitor had been vehement in stating that humans should never touch wild animals because it's bad for them, it habituates them, and that could make them less competent in the wild. Terry reported that he'd asked her if she had touched a whale. "Yes," she had said. And "yes" say I, for when we touch each other we are both changed by that, but it doesn't necessarily ruin us. I am inclined to question the reasoning that always has humans as agents and other animals as objects affected by the agentic humans. I have come to think that we and they are agents interacting. Caution is good, but so is contact, so is care, so is the awakening in each that the other brings.

The next day Anneliese got her wish. We went out again in choppy waters and again saw whales, mostly mothers with toddlers, swimming at a distance, parallel to the sandy shore and especially abundant at the mouth of the lagoon, close to where the inlet meets the Pacific Ocean. The whale kids were moving into tougher lessons now. There were some beautiful breaches, when adult whales jump entirely out of the water, most often seen as a fluke returning to the sea at a sharp angle. It takes very good luck to see a complete breach; you have to happen to be looking in the right direction.

Returning from our island lunch, the waters were preternaturally calm, so calm that shades of white and of blue were distinct, so calm that we could see the birds overhead reflected on the still water. Ranulfo motored us slowly and quietly along. There was an odd absence of the usual playtime parade of mother and baby whales approaching the boat. And then two of them emerged, a very large mother whale who was herself interested in the boat. I wanted to bolt up and get to her as she approached, hoping to have the coveted moment of eye-gazing. I was sitting in the back on the starboard side, and I had to remind myself to resist my eager impulse and to get out of the way so that Anneliese would have plenty of room to maneuver. She'd come a long way to touch a mother whale and the moment was here.

I decided I'd try to get a photo for her. I exerted enough effort to get a good angle that I tripped backwards over the seat and landed legs up in the back of the boat. Embarrassed, I quickly got myself upright and got my camera in front of my face just as Anneliese stroked the giant alongside us, praising her and thanking her for the visit. The photos were more of a revelation than I hoped for, shot through with light at the place where human hand and whale connected. The baby, who had stayed uncharacteristically on the other side of mom, came for an extended visit just as mom wanted to leave. I petted his head, first on one side of the boat, then the other. And then Ranulfo motored off in support of mom.

And so go the days, quiet sunrise with a cup of coffee on the beach, breakfast with a group of interesting strangers, mornings looking for whales, lunch on the sandy beach of an uninhabited island strewn with signs of sea life, current and recent, early afternoons with playful and curious little whales approaching the panga, monitored by their mothers, with plenty of opportunity to pet the little ones. The big whales dive deeper than do the babies and they swim under the boat regularly. You can be fully engaged watching a whale, maybe interacting with it, and suddenly it's gone. You scan the water but see only the round footprint where it was, or sometimes even a trail of footprints that lead off to where you will see it next. But often the whale has simply dived under the boat to get to the other side. I have always sensed when the whale is under the boat and often know where to expect it on the other side. I do not think this is a matter of some special intuitive attunement to the whales, but rather a sensory awareness at a more subtle level than is common in daily life on the ground. Part of the fun for me is sensing where the whale is going, and going there to meet her. The best hours of the day are spent learning to see and to know a very different kind of being, and so also to draw out different capacities from within yourself.

If you're lucky, you may get the opportunity to rub the whale's baleen, something that seems to be a particular pleasure for them, rather like us humans having our necks massaged. The whale is swimming alongside the boat, angles its head toward you, gently opens its mouth so that you see the baleen. If you're feeling brave, as I most often and sometimes foolishly am, you put your hand inside the whale's mouth. If you are going to do a decent baleen scrub, you put your forearm inside the whale's mouth because it is a very big mouth. I can't help but have at least that one moment when I imagine what might happen if the whale hiccupped. But the whale never does, you are never crushed or dragged out to sea, much less maliciously carried under and drowned. Rather, the whales treat us with the utmost gentle care. I think they are more careful than I am when I carry a spider out of the house, more patient and calm, less twitchy.

On the evening of Anneliese's wish-granting we shared another glass of wine, as was the habit of our short acquaintance, and soon the group was gathered round, waiting for dinner and wondering together the oft-wondered question: Why is this happening here? It is so unusual! We had all heard the very accurate warning never to get between a wild animal mother and her baby, because wild mothers are notoriously fierce in protecting their offspring. As scientist Toni Frohoff, a specialist in whale stress and well-being, said in an interview, "It's extraordinary. At precisely the time when you'd expect them to be the most defensive, they're incredibly social. They'll come right up to boats, let people touch their faces, give them massages, rub their mouths and tongues."[13]

After dinner, during the whale talk, the pondering of this question continued: What is going on here? And suddenly it occurred to me that, in addition to being in the company of creatures so big that they are predisposed to being relaxed, we are in a social atmosphere heavily laden with oxytocin, the "trust" hormone that all mammals have and that we experience most intensely in our mother-infant bonds. We were all floating in a virtual sea of oxytocin!

I cannot begin to comprehend some aspects of being a whale—like needing to breathe by conscious choice or risk drowning, like living in water rather than on land, and in a body filled with blubber on a frame of hollow bones, a body that is effectively all ears, entirely tuned to sound waves from head to fluke. Nor can I imagine all the issues related to matters of size, like the six-thousand-mile commute and back. But oxytocin? Yes! The affective system that Jaak Panksepp labels as care is the filter through which I—sitting in the mess tent, with the solar lights glowing overhead and with the whale posters on the walls—am suddenly looking at the phenomenon of whale and human communion here in the Baja lagoons.

The hormone oxytocin floods new mothers and their infants, allowing them to be fascinated by each other, to want to be together, to prefer that other's face and smell and mannerisms to all others.[14] If it sounds a lot like being in love, that's because it is a lot like being in love. Developmental neuroscientist Allan Schore describes it this way: "Dyadically resonating, mirroring gaze transactions thus induce a psychobiologically attuned, affect generating merger state in which a match occurs between the accelerating, rewarding, positively hedonic internal states in both partners."[15] That is to say, when we gaze into each other's eyes, we are actually biologically tuning to each other and it feels really good; and when I see it feels good for you, it feels even better for me. That kind of tuning to another first happens between mothers and babies and, for all mammals, it is the platform on which the infant's social development rests. The calming presence of the mother and the needing presence of the infant release oxytocin in the two of them, setting the threshold for feelings of trust and threat that will guide this infant as he or she grows and encounters others in a bigger social world. The hormone creates a state of resonance between two beings that allows not only for them to fascinate each other, but for others to be drawn into the fascination, stimulated by their own release of oxytocin. While it is strongest in the mother-infant bond, it exists in all bonded relationships and is stimulated by experiences like touch, eye-gazing, and sexual contact, and then even by the image or memory of these.

When I taught college psychology, I used to joke that we should put oxytocin in the drinking water. Now I suddenly perceive myself and these others drawn here from various distant shores as floating in a sea of oxytocin, protected indeed from the world out there, a world that cannot be so readily trusted. Perhaps it is the oxytocin that allows for this uncommon trust across species boundaries. Yes, the whales were nearly extinguished here and some of them no doubt remember that, so we might expect their trust in us to be challenged. But we are small, and the many of us in a single boat could easily die from the slightest carelessness on the part of the whales, a carelessness that simply never happens. In short, there is care, not only in the mother whales for their babies, but in our care for them, facilitated by nothing short of the surprising possibility of their care for us.

These thoughts pulled me up short and set me down in a different place, with a different perspective. I took them into my dreams and my waking desire on the last of my days at Laguna San Ignacio. The women from Minnesota had decided at the end of the second day that they had seen enough of the whales. I suspect that some of them had had enough of roughing it. But what I know is that they said they were ready for showers and beds and a chance to see the blue whales, the biggest animals on Earth, in the Sea of Cortez. Though it seems in one way unimaginable to voluntarily leave the baby whales here, I too might jump at the chance to see the blue whale again, the largest animal ever, the animal whose very, very long body is adapted to create homeostasis in three different atmospheric pressure zones simultaneously while bobbing vertically in the water. The departure of four of our company allowed for our final days together to be calm, and for us to engage in deeper conversations, to have quiet times alone and quiet times together.

On our last morning, I allowed as I had not looked into the eye of an adult whale, though I had petted many a baby, scrubbed an adult baleen, and gotten a face full of blow, that musty-scented exhalation of a whale through its blowholes. I still yearned to look into the eye of an adult as I had looked into the eye of Barbilla. Sarah was fairly rapturous with her newfound sense of being not just in the same world as these others, but in a real relationship with them. And I had yet to avail myself of the promised one-to-one conversation with Ranulfo.

The morning on the lagoon was quieter than usual. The whales were in serious preparation for their upcoming long trek and we deduced that they were therefore less inclined to dote on us. On the beach of the island where we stopped for lunch, I sat with my back against a rock, talking with Ranulfo. I had watched him spear a fish under a rock just at the shore. "Dinner," he said. "Like the wild birds, I know where to get food." I invited him to join me and told him I'd like to ask him some questions. Still enamored of my oxytocin hypothesis, I asked him why he thought the whales were interested in us, why they let us touch them, why the whole scene here. He replied that the gray whales are friendly and social by nature, that their natural trust is probably part of the reason they were so readily slaughtered in the past.

He told me about his extended family and their work in continuing his father Pachico's legacy, protecting the lagoon and the whales. He told me it was a joy to see the first whales return each year. I asked if he took the summers off. I don't remember his exact words but I remember my impression: "Hell, no!" When he wasn't taking people out to meet the whales, he was fishing, as his family had for generations. I asked if he had any time off and he said that he spent a bit of time in late autumn making his drawings of the whales, which he sells. I pressed on the question about rest and he replied that he and his family rested the way wild animals do, when the weather required it, and he offered that it seemed to work well for them.

We returned to the boat well-fed as usual, knowing this to be our last two hours with the whales. Knowing, too, how hard it is to get to them, we were a bit subdued. But soon I was standing in the back of the boat, again whistling for whales. And soon enough we were approached by a very playful and social mother-and-baby pair. Sarah and I were on the starboard side, me in the back, she in the front. Ralph and Anneliese were on the port side, all of us with plenty of room to move, and feeling emotionally expansive too. The adult whale came up near Sarah for an intensive visit. She petted the whale and talked to it, as did I. And when the whale began to swim away, Sarah said, "Mamma whale, one more thing. Anne would really like to look you in the eye. Can you do that?" And to everyone's amazement, the whale lifted her head from the water directly in front of me and before I even knew what was happening I had reflexively taken a photo of the eye of the whale before I dropped my camera into the boat (also without thinking!) in order to have that moment of eye to eye, the two of us bathed in the oxytocin gaze.

The whale-shaped piece of deep-blue glass rests gently in my hand now, and at once I feel again that gaze.

Riding in the van back to our hotel in Loreto, Anneliese continued to stitch together the story of her life. She was born just outside of Köln during World War II. Her father had been exempt from military service because of his age and the fact that he had already fulfilled his military service in World War I. He was terribly relieved, she reported, because he detested Hitler, and was glad not to have to put his life on the line again, this time in protest. But on very short notice, her mother, pregnant with her, was ordered to stay with a host family on a farm in the countryside as part of the *Kinderlandver-schickung*, a program to remove women and children from places especially vulnerable to air raids. Her mother expected Anneliese's father to join them the next day, but he never came. After Anneliese was born on the farm, her mother walked the thirty kilometers back to their home, carrying her infant, only to learn that the child's father had been killed in an aerial bombing on the very night that she had left, in the building where they had lived. Having nowhere else to go, she returned, again by foot and carrying her infant daughter, to the farm of the host family, who were no longer required to provide hospitality. But they agreed that Anneliese and her mother should stay there, in exchange for work. And so, Anneliese told me, she was placed in the barn with the sheep and the cows and the chickens every morning as her mother went off to work the farm, and retrieved from their care every evening. She told me that she thought her bond to other animals began right then, in her infancy among them. I quietly considered, and still do, the role of oxytocin.

Two things had jumped out at me from the waters of Laguna San Ignacio. First, the confirmation of my hypothesis that we have some things in common with whales that provide a basis for cross-species empathy, oxytocin and the primary affect (or instinct) of care being one of them, and developmental trajectories being another. We all move from infant to toddler, to child, to adolescent, to adult, to older adult, if so fortunate, realizing the potentials of both our species and individual natures as we traverse our time on Earth. And we can understand a great deal about one another's lives on the basis of these two things. The second thing that stood out for me was the experience of a rather radical shift in perspective, from the commonly held idea that we should save the whales to the clear notion that it's not just what we do to and for the other animals that matters, but also what they do to and for us, and who they are in their own worlds.

Have the whales come to teach us forgiveness? Or, perhaps, are they so gregarious in their natures that they can't help but play with us? Are the whales domesticating us for their purposes? Do they want us to love them so that we won't destroy them again? Do the whales have a "save the humans" campaign, in which they try to teach us not to feel so separated from the rest of the world but to feel instead that we are part of the world, and even perhaps that the world cares for us?

In truth, we really don't know what they are doing with those excellent big brains of theirs.

CHAPTER 7

Why Cats, Florence?

During the Perseid meteor showers of 2018, a slender young brown tabby cat brought her two tiny kittens to live in the big olive tree behind our house. We named the mamma cat Stella and her two kittens Alpha and Centauri. Alpha looked a lot like her mom, and Centaura (as she became when we learned that she was a girl) looked like nothing I'd ever seen, black with bits of orange poking through, with short legs, big feet, an extravagantly puffy tail, and a face divided down the middle of her nose, black on one side, blazing orange on the other, with huge yellow eyes. Though I have long been partial to brown tabby cats, it was Centaura who captured my heart and who still creates the gravitational center of my cat lessons. Who knew that the small feline, *Felis catus*, could have such power? Well, it turns out that many people know the power of the house cat and, though I grew up with both cats and dogs, I am a latecomer to learning the magic of cats.

Stella was hungry when we met her, nursing two kittens on whatever she was able to hunt in the fields while teaching and protecting her babies. So we began to set out bowls of food for her, taken from the supply of our indoor cat, JouJou. Stella was wary and stealthy in her approach to our offerings of food, but deeply appreciative, and after some weeks she began to bring her kittens with her to eat the kibble. We had in mind to catch Stella and to bring her to our vet to be spayed, but Stella got ahead of us. She left her kittens, now trained in hunting, avoiding danger, and locating bowls of kibble. She slipped back into the *campagna*, the countryside of southern Italy from which she had emerged.

When we began a house remodeling project in October of that year, at the same time that the weather changed to rainy and cold, we lost the safe place, the open top of the wide stone garden wall, where we had been feeding the cats. After we had torn apart that area, I called for the kittens but couldn't locate them. I left food for them in places that seemed relatively warm and dry. Gradually they began to reveal themselves, sheltering from the storms under a long porcelain sink that was turned upside down on the ground, awaiting installation, or on top of the old tractor half overgrown with bamboo. When I mewed for them, they mewed to let me know where they were. I was amazed that "my" kittens answered me in this way. But I have since learned that cats mew specifically for humans; it's not how they communicate with one another, but they know that we need sounds and "cat words."[1]

Though they became increasingly tame, the kittens continued to hunt birds, rodents, lizards, and insects as their mother had taught them to do. I had seen Stella bring captured prey to them, setting it down and backing away to let them figure out the eating, and later the hunting as well; all so distinctively different from a dog mom who regurgitates masticated food directly into the mouths of her puppies. These were working kittens, and our once significant rodent population was already vastly reduced. But the kittens increasingly relied on our kibble so they hunted less as the months of their first year unfolded. I began to suspect they would move in to share a comfortable old age with us one day, unless the cycle of life claimed them first.

The wild kittens were and always will be much more wily and skilled than is our house cat, JouJou—she who watched in amazement every day as the feral cats climbed high into trees, leaping from limb to limb, until one day she steeled herself and climbed a tree. Her small victory was one of the most touching animal events I have ever witnessed. It was scary for her but, though less athletically gifted than her wild cousins, she overcame her fear and leapt. Then she looked down from the tree in a kind of tremulous victory grin. Now she jumps from the garden wall to the high branches regularly, if not quite as gracefully as the feral cats. And she mews at the door to be let in and out. Alpha and Centaura, for their parts, watched amazed as JouJou allowed herself to be picked up and cuddled, until one day they each allowed themselves to be briefly picked up and petted, if not yet cradled and carried to the nearest soft cat bed. The big-footed and rib-thin feral cats had become a bit heftier, though not much, and JouJou the tree climber had become much lighter on her feet and more confident.

One day as I sat in my studio an acre away from the former feeding station, working with the windows open, I heard a loud mewing at the window. It was Centaura, asking for food after some big storms had kept us in our separate shelters for a couple of days. I was overjoyed, and the serious courtship commenced. Centaura began to call me and to respond with loud mews, even from distant fields, when I called for her. One day she screamed for me, and I ran to find her treed by a pack of yelping and salivating feral dogs. I chased them away and she waited until they were really gone before darting into a pile of rocks where, no doubt, cautious Alpha waited for her. Invariably, the sisters were together and they arrived at mealtimes together.

These feral cats allowed themselves to be tamed incrementally, a little bit closer each day, with us looking out in the same direction together, not directly at one another—yes, with a bit of fish on the ground between us—until trust was established. But it was Centaura who was interested in the relationship, Centaura who was curious about me, Centaura who was playful and demanding at the same time. In fact, I will never forget my first touch of these cats. No surprise, it was Centaura. I called her from the olive trees and I left a trail of oily salmon bits that led from one of her shelters to where I sat, twenty feet or so away on a tree stump near the woodpile where she liked to hide. I was wearing rubber garden clogs and the last bit of oily salmon sat on my toes. She started eating the fish bit that was furthest away, eating her way closer and closer, wary when she got to my feet, but finally she came and took the bait. I slowly held out my hand, not too close. She backed away, then returned to eat the rest. Having decided to take the risk, she arched her back up into my extended hand and let me pet her while she ate, and she purred mightily. Oh my heart, my dog-centered heart. I was changed right there by this little wild fairy of a cat. It would be a while before I got that freedom of touch and that amazing purr again, but it was more than enough to keep me interested. As for her, she had marked me as her own with the scent glands in her arched back.

It was also Centaura who wanted to come into the studio, and she let me know by hanging around, coming close enough to look inside. I left the door open. She entered and sat very still just inside the doorway, methodically looking the place over, memorizing its contours, walls, ceilings, corners, the tops of furniture, the undersides of furniture. Then she went under the bed and sat silently on my wool slippers while I worked. When I got up, after two hours or so, she left. The next time she came in she jumped up on the desk and leapt to the top of the wardrobe. She smelled the dog crate and then went inside it to have a nap in the soft warm of it. The third time she came to visit, I did not know she was there. I left for a break and, when I returned, she had urinated on the bed. I understood that she had just offered to put a ring on it, but I had to get sheets into the wash and get to an appointment. So I tossed her out as gently as I could and got on with my tasks. She stayed away for a few days. I imagined her feelings were hurt at the seeming rebuff. But when she returned, the courtship resumed as though uninterrupted. Alpha was most often nearby, but she was always more utilitarian than her adventurous, curious, and seductive sister. Alpha, having established her territory and food source at our little house, never left. But Centaura, so deeply and appealingly adventurous and curious, left for days at a time, days when I would walk and drive the campagna roads calling her, looking for her, pleading with the powers that she not be poisoned or killed by a pack of feral dogs.

One day she returned after four days away with a deeply embedded foxtail weed and a suppurating wound on her back. She let me clean and dress it, but the infection returned within hours. She let us crate her and take her in for surgery, and she wore the cone and lived in our bathroom for ten days, all without drama. She was curious, exploring every cranny and asking for lots of loving attention. After she was completely healed, she joyfully bounded back into the campagna, a cat changed in her comfort with humans.

Centaura, more than any other animal whom I have personally known, exemplified the delight of the most fundamental of the affective systems described by Panksepp and Biven, the seeking system in the brain.[2] The seeking system in the mammalian brain was confounded with the reward system at first, but these are actually two distinct systems. The reward system is facilitated by endogenous opiates, endorphins, and is connected with the experience of sensory pleasure or satisfaction of desire. The seeking system is facilitated by dopamine. Both systems create pleasurable sensations, but the dopamine seeking system derives pleasure from the brain itself, the pleasure of desire itself, which rewards animals for wanting to explore the world around them. In other words, our ancestors, through the mechanisms of genetics, have molded us in such a way that we are rewarded for having the impulse to move beyond ourselves, to get up and go, to encounter the other. Panksepp and Biven describe the system succinctly and distinctly in this way: "When the SEEKING [sic] system is aroused, animals exhibit an intense, enthused curiosity about the world."[3] This system provides the pleasure of anticipation, the feeling of arousal about what might be discovered or experienced. The pleasure of getting dressed for an outing, or of getting undressed if the outing is a date, or of imagining revenge on a rival, or of experiencing hunger pangs as a consequence of good-smelling food: all of these are examples of seeking-system pleasures.

Centaura and I shared a bent in the direction of the seeking system, and I think we recognized this in one another across our species boundaries. For me the arousal pleasures of exploration, adventure, and anticipation almost always outdo the pleasures of satisfaction, and I am sure that was true for Centaura too. Even her play was distinctly robust and creative—she made for herself a game of jumping from the top of the garden wall to the boat-shaped metal frame of the hammock stand, grasping the top with all four paws, then letting gravity slide her down the long curved pole. She was driven to explore and to master her world, and she was almost always smiling.

But came the day, just after the Perseid showers of her first birthday, when Centaura left and never returned. The last time I saw her, I was striding across the field to my studio intent on my work. She bounded along next to me, rolling on the ground to show me her precious and vulnerable tummy, inviting me to play. I noted how long her stubby legs had grown, now seeming to match her giant feet. I looked at her and thought, we are buddies now, we have all the time in the world to play. But we didn't have all the time in the world. In retrospect, I think a fox got her, and I may have even heard it taking place. At about three in the morning, I recognized the sounds of a hunt, and I thought it was a hedgehog squeal. A few days later I discovered the foxhole under the garden wall. I should have kept her in. I should have played with her that morning when I chose to work instead. I should have noticed the foxhole and closed it. I should have gotten up to see what the hunt was about, even though it didn't sound remotely like a cat to me. Next time. Next time. But it won't be Centaura, my fairy cat, next time. Alpha, the careful cat, less exuberant than Centaura, comforts me still. Centaura has gone the way of most feral cats, to an early death, but after a wonderfully adventurous short life. Alpha, with her human relations to help her and her cautious temperament, may well live a longer and more sheltered life with us.

I used to wonder how cats, seemingly solitary and independent, evolved into the creatures who appear in our social media posts. What did we see in each other? What drew us together? It turns out that the latter question comes first. When we started to farm and to have grain bins, the mice came in droves to the grain and the cats came in droves to hunt the mice.[4] This territorial arrangement created evolutionary pressure on cats to extend their social repertoires, putting our two species in proximity, and both species benefited from affectionate bonds. It also put cats in relationship to one another, often a more stressful situation for them.[5] Because cats offer only a singular skill to be exploited by humans, that of hunting rodents, "to make the leap into domestication, it [the cat] almost certainly had to become an object of affection as well as utility."[6]

Anthropologist Elizabeth Marshall Thomas says of the reputation cats have for being antisocial, "It is my impression that cats seem unsocial to us only because we aren't good at recognizing the signals of other species. We interpret cat signals to tell us, for instance, that they don't care about us and don't miss us when we're gone. If people were giving off similar signals, we'd be right. But they're not people, and we're wrong."[7] So when the cats came to the human grain bins to hunt the mice, they claimed their territories with urine, as cats do, and they stayed in their territory where, from their point of view, we happened to be. And then we and they grew to like each other enough to hang out together for the next ten thousand years.

Though we don't understand many of the underlying mechanisms, the positive effects of companion animals on human health have long been noted, both in classical medical texts and in contemporary research. After listing the beneficial effects of fresh air, sunlight, and cleanliness, Florence Nightingale gave this advice in her 1859 text *Notes on Nursing*: "A small pet animal is often an excellent companion for the sick, for long chronic cases especially."[8] During her lifetime Nightingale kept more than sixty cats, reputedly up to seventeen at a time, and in her old age she surrounded herself with them.

Why cats, Florence? Florence can't answer me now, and I don't know what she'd say. But I think it may be that the effects of cat purring make them especially healing animals for humans. Some cat experts think that cats purr to communicate, and primarily to communicate contentment. In fact, Elizabeth Marshall Thomas made the bold claim that "no cat purrs unless someone is around to listen."[9] I wondered how she knew this, that cats are not purring when she is not around? It's likely, I thought, that we notice when cats seem to be telling us how satisfactory they find us to be, but we are less able to know when they purr for themselves. Now the research has been conducted with hidden cameras to learn just what cats are doing when we are not there. It took decades to figure out the physiological mechanism of purring, vibration of the vocal cords, and we have learned that there is no singular and simple explanation for this cat superpower. Cats purr in many ways and for many reasons.

"Yet they also purr when we're not around," says Stephen Dowling, referencing a conversation with Marjan Debevere, a cat psychologist in London.

Jason among the pine trees. PHOTO BY ANNE BENVENUTI.

Lioness, camouflaged. PHOTO BY ANNE BENVENUTI.

Benedict Goat, back on his wheels after falling down. PHOTO BY ANNE BENVENUTI.

Bill Kauffman and
Sebastian Pig, in a
cuddle nap after lunch.
PHOTO BY ANNE
BENVENUTI.

The magical world of blue hippos. PHOTO BY ANNE BENVENUTI.

Lions, after lust. PHOTO BY ANNE BENVENUTI.

Baby elephant play and maternal discipline. PHOTO BY ANNE BENVENUTI.

Mother and baby cheetah. PHOTO BY ANNE BENVENUTI.

Eye of the whale. PHOTO BY ANNE BENVENUTI.

Centaura, the fairy kitten. PHOTO BY ELIZABETH HILL.

Loki, Siberian fox. PHOTO BY THOM ESPINOSA.

Bees, a mammal in many bodies, at sunset. PHOTO BY ANNE BENVENUTI.

"It's likely that purring has communication, appeasement, and healing properties," adds Gary Weitzman, a veterinarian and chief executive officer of the San Diego Humane Society.[10] Cats purr when they are stressed, when they are sick, when they are in pain, when they are dying, as well as when they are content. I find it hard to imagine that a cat in pain would purr only for the purpose of letting his human know that he is having a feeling. While many people find the purring of cats to be soothing, and while it does most often seem to communicate the cat's contentment, purring may also be a form of both physical and psychological self-healing for cats. "Cats purr during both inhalation and exhalation with a consistent pattern and frequency between 25 and 150 Hertz. Various investigators have shown that sound frequencies in this range can improve bone density and promote healing."[11] Other investigators think that different frequencies promote healing in soft tissues.[12] This didgeridoo type of breath work, creating continuous vibrations on both the inhalation and the exhalation, may help to explain the relative absence of orthopedic injuries that we would expect to find in such an athletic species, the kinds of injuries that we do in fact find at greater rates in dogs and humans.

Cats purr, I think, to communicate both desire and contentment, but also to soothe, self-regulate, and physically heal themselves. And I further hypothesize that because they know it helps, they selectively offer it to other animals, like humans, as both an expression of contentment and as a gift. No, I don't suppose that cats sit around engaging in abstract thoughts like, "My human is hurting. When I am hurting, purring helps, so I will purr for her because I think it might help her." Rather, it is the kind of basic body knowledge and communication that we share across species. Simply put, we know and they know that purring is good; it's good to help you feel better, and it's good to express feeling good. It reminds me vaguely but definitely of the role of soft jazz music in my life. I use it when I am stressed to create contentment and when I am content to express that contentment. But cats make their own jazz and cats are probably actually healing their own bodies when they do so, and maybe cats are also sharing the gifts of contentment and physical and psychological healing when they purr for the people they love.

Whoa, there! The people they love? Everyone knows that cats don't really love their people, but simply require the services of a human and devise clever systems of manipulation to get what they want. Not so. In the words of cat researcher and scholar John Bradshaw, "Cats are social animals too; otherwise, they could never have become pets as well as hunters."[13] Essentially, of course, all mammals are social animals, with small variations in the oxytocin bonding system resulting in species differences in degree and style of sociality. For cats, sociality seems to be more of an option than a necessity, as is easily seen in the comparison to dogs: "Whereas most dogs are inclined to bond strongly with their people, if properly socialised with humans early in life, this is more of an option for cats, making the strength of bonding and attachment styles with humans more variable in cats than they are in dogs."[14] Cats, it seems, are very flexible animals in their expression of sociality: "Due to the great range of bonding and interaction styles between cats and people, these intraspecific companionships may be more variable and, therefore, individually unique than the relationships people maintain with any other companion animal species."[15] In other words, humans and cats both have ranges and degrees of sociality, unlike dogs, who are necessarily social pack animals, resulting in many possible combinations of human-cat relations. The human-cat dyad allows for greater expression of individual differences, times two.

But how can we know with certainty that real relationships exist between humans and cats? How can we be sure that we are not simply projecting our human desires onto cats and then seeing their behavior through our human lenses? One study that elegantly addresses this question breaks down the question of meaningful relationship into minute patterns of behavior used in relationships by which each partner influences the other, in continuous information exchanges. The researchers hypothesized that "human-cat dyads may be regarded as long-term valuable relationships . . . probably characterized by dynamic negotiations of interests between partners."[16] In their study, they analyzed microbehaviors, exchanges of various kinds of signals given and received to mutually modify the behavior of the other, exchanges by which cats (in this study, domestic) and their people negotiate the meeting of needs, feline and human. In other words, they were examining the behaviors of the human as well as the cat in each dyad.

Perhaps not surprisingly, they noticed the influence of personality and temperament factors. If either the human or the cat scored high on neuroticism or extraversion, for example, it would shape the interactions and behaviors between the pair. Human personality traits of extraversion and conscientiousness were related to higher interaction patterns, and the human trait of neuroticism (very roughly, neediness) was, again not surprisingly, negatively related to the number of human-cat interaction patterns per minute. But the chief finding had to do with gender, not of the cat but of the human. The female humans in the study were more active toward their cats, and these cats approached their person more often than did the cats whose person was male. The researchers concluded that female owners tend to have a more intense relationship with their cats, a finding which corresponds to historical material on the association of women and cats dating back to the Egyptian goddesses Bastet and Isis, and to Diana in southern Europe.[17] As for cat variables, older cats were found to engage in less complex interactions with their humans, reminding me of the mealtime silence of some human couples married to each other for many years. More active cats had fewer uninterrupted exchanges per minute. The sex of the cat was irrelevant to their interactions with their human partners.

A 2019 study conducted at the University of Oregon focused on attachment patterns of cats in relation to humans, using the strongly established model and research protocols of attachment theory. Interestingly, this study found that attachment style in cats is similar to attachment style in humans and dogs, with cats even showing slightly higher levels of secure attachment than dogs.[18] It transpires that a cat has a secure attachment to their person in the same proportion as human children have secure attachments to their parents, and slightly more than those of dogs to people. Dogs have slightly higher rates of anxious attachment, likely because they are oriented to membership in a pack and so are less independent than cats. Dogs need to please; cats don't. In Bradshaw's words, "Cats somehow manage to be simultaneously affectionate and self-reliant."[19]

Feral cats like Alpha and Centaura represent a kind of middle ground between wild and companion animals, living on the periphery of human society but not entirely dependent on it. Their domestic counterparts, house cats, are members of human households and actually, I think, working animals too. House cats provide services to humans, but the nature of the service has changed from mouse hunting in the grain bins to performance of a kind of healing work that falls somewhere between psychotherapy and nursing. Feral cats sometimes provide the service of rodent control, but their hunting is increasingly seen as an ecological threat to their prey species, especially songbirds and small reptiles . . . though no one seems to worry too much about the fate of the mice, rats, or insects.

I discovered, while reflecting on what I knew and didn't know about feral cats, that I knew two people who had devoted significant time to feral cat care. Sabina Magliocco and Mary Jean Kraybill, a professor and a business administrator, respectively, are two of many who have been involved in the founding of trap-neuter-return (TNR) programs in American cities. Sabina's project at California State University, Northridge (CSUN) was developed to deal with a specific set of problems in a carefully defined legal jurisdiction

and geographical setting. Hyde Park Cats, Mary Jean's group in the neighborhood of the University of Chicago, arose organically out of the intersecting
needs of people relative to their cats in a large urban neighborhood, and it
remains less defined in terms of its goals and methods, even its territory.

When I first interviewed Sabina Magliocco, professor of anthropology at
CSUN, at her former home in Los Angeles, I also met her cats: Callie, the shy
feral who had come to her as a very young foster cat with a broken jaw, and
Mac, then a foster cat and now a permanent resident. The two cats were
holding down opposite ends of a window box with a view of a Mediterranean garden, and Sabina wondered aloud if they would ever become friends.
The cats stayed put in their relative places while Sabina cooked a delicious
frittata, telling stories of the feral cats on campus at the same time.

I was surprised by how casually Sabina described a TNR program as part
of "maintaining the campus feral cat colonies."

"Is this a normal thing, that college campuses have feral cat colonies?"

Oh yes, she assured me. Not only college campuses, but hospitals, parks,
and any other urban or suburban settings that have open grounds and trash
cans are likely to have feral cat populations. TNR has largely, if sometimes
controversially, become the replacement protocol for the earlier practice of
capturing and euthanizing feral cats.[20]

Because she had just been in Sardinia doing anthropological field work
with her students, Sabina brought the second interview about her work with
feral cats to me at my home in southern Italy. We sat in the summer sun on
the veranda with Alpha and Centaura strutting and stretching nearby as Sabina reflected on her involvement in bringing creative animal care and humane education to her former workplace. Sabina gently reached a hand in
Centaura's direction, making a soft purr sound, but Centaura did not take
the bait. Sabina told me that soon after she had arrived at CSUN she had
discovered feral cat populations. Part of the legacy she left when she moved
to the University of British Columbia was the establishment of CSUN Cat
People, a university-recognized group organized to care for the feral cats on
campus, and to educate the human campus community about their respon-
sibilities to and for feral domestic animals.

"Where do the cats come from?" I asked. One obvious answer is that
they come from other feral cats, as kittens, hence the desirability of TNR
programs. It stands to reason that feral cats can reproduce faster than they
can be exterminated, so that, even for humans with no concern for the cats,
euthanizing feral cat colonies is not an effective solution. The Southern

California–based organization Best Friends Catnippers explains it this way: "Euthanasia has proven to be ineffective in controlling feral populations because the remaining animals reproduce more to fill the territory vacated by those euthanized. It is also not effective from a cost standpoint, consuming far greater resources than the TNR approach."[21] To trap, neuter, and return, by contrast, involves members of the community humanely transporting feral cats to a vet who will spay or neuter them at subsidized cost, and then returning them to their colonies where they can receive ongoing care and feeding.

Sabina described other sources of cats specific to the university community. Students might acquire a pet cat and then at some point move on, releasing the cat on campus to fend for itself. Other students might think the campus a good and safe place for the release of animals no longer wanted at home. And people living near the campus seemed to have shared in the same notion, that it was a good environment for the release of unwanted animals, who would, after all, be fed. Sabina offered that the cats themselves sometimes ran away from homes—for example, in the aftermath of the 1994 Northridge earthquake—and that others simply preferred the campus with its feral colonies to their human households. In these various ways, the feral cat population on the CSUN campus had come to be somewhere between seventy-five and eighty-five cats when Sabina arrived as a new professor.

Having grown up with cats and worked in wildlife rehabilitation, Sabina knew that she could not just start feeding the hungry cats without creating new kinds of problems. She knew she would need to trap, neuter, and return the feral cats so that they could continue to live, but could not continue to reproduce. And, because she would be feeding them, their hunting behavior would be vastly reduced. Feral cats that can't reproduce, she noted, actually serve to keep their environs rodent-free, and they live better lives for being fewer. Feral cats that do reproduce in urban settings can be the source of a snarled knot of problems, from threatening songbird populations to spreading disease among both house cats and other feral cats, to increasing chances of starvation for each of the cats competing for environmental resources.

The situation of the feral cats at CSUN was politically sensitive. Some members of the university community had been quietly feeding the cats. Others were irritated by the trash, paper plates, and takeout containers blown about by the wind. Still others had jobs that involved the capture and euthanizing of the problematic cats. Employees who were working in temporary buildings in the wake of the 1994 earthquake protested that the cats who were living under the skirting of the buildings in the crawl space were infusing the buildings with fleas. And some cats found their way to the day-care center's playground sandboxes, which they then designated as family-sized litter boxes. The people who had been feeding the cats did not want attention drawn to the cats or to the feeding of them because they feared such attention would cause the deaths of cats they'd come to love. Others did not want to sanction the feeding of cats, thinking this would only increase the population, both by reproduction and by affirming the idea that the campus was a great place to abandon animals with minimal guilt.

Looking to borrow crates for trapping feral cats at home, Sabina contacted Best Friends Catnippers, a local organization that provides low-cost spay and neuter services. In the course of her conversation with Catnippers, Sabina brought up the subject of the campus cats and learned that a graduate student, Jessica Taylor, had already contacted them about the very same cats. Together, professor and graduate student decided to address the issues. They determined that they had two initial tasks: to convince the cat feeders that involving the university administration was a good idea, and to convince the administration that feeding the cats would make TNR possible, thus reducing the population of feral cats, resulting in a cleaner and safer campus. The initial goal of CSUN Cat People was to get to zero feral cat population through humane methods, though that goal was later revised when the campus community came to understand that it wanted a small colony of feral cats to control the rodent population.

Happily, Sabina discovered that the stepdaughter of CSUN's president, Jolene Koester, had just set up a TNR program at a college campus in Chicago, and due to this synchronicity her request that the administration get involved was warmly received. With support from her university's president, Sabina began to work with Tom Brown, director of the university's physical plant, offering for his review successful programs of TNR on other college campuses. Soon Tom had the campus grounds crews developing designs for feeding stations, within the parameters that the stations had to be resistant to other kinds of animals and unobtrusive, blending with the environment. The cat stations also had to be clean and cleanable. With trial and error, Sabina told me, they landed on a clever design "using vinyl fake rocks that you can get for about sixty bucks at a home store—and then cut two holes in them so that the cats could come and go. They created cement bases for them and attached them in a way that you could lift and feed, but also lift and clean, just go in with a hose and clean the whole thing." The fake rock design worked for the cats and for the people. "Within about a year or year and a half we had twenty feeding stations around campus, and we were trapping a lot of cats."

Sabina, the ethnographer experienced with interviews who had earlier described herself as being like a race car moving fast on high octane fuel, steered the interview deftly and moved it quickly. But as I pictured the trapping of cats in "have a heart" crates, in a specified location over a period of months, I had a burning question. I interrupted to ask how they knew a sterile from a nonsterilized cat.

"While they are anaesthetized, they clip the top of one ear so that you can tell a neutered from an unneutered cat at a glance. One of the things we did was to make an inventory, so that we would know and be able to report the effects of the program. So we know that within about a year we went from seventy-five to eighty-five cats down to half that number. Some died from disease, or were killed by coyotes, or hit by cars. Some of the volunteers took cats home. There were several factors—lack of reproduction, death in the normal ways that feral cats die, adoption—but within two years the population was halved."

Mary Jean Kraybill told me similar stories of Chicago, where the feral cat issues are much the same as they are in Los Angeles, except that the winters are bitterly cold, necessitating insulated and warm safe spots, and the neighborhoods are densely urban. And around Hyde Park, one of the issues with caring for the cat population is that in some neighborhoods the people who love the cats can't afford to feed or neuter them.

When I entered Mary Jean's Hyde Park apartment for the first time, I noticed affectionately the classic visuals of the University of Chicago neighborhood: antique chairs of mix-and-match vintage covered with cushions, books, more books, art prints, and, of course, a bit of cat hair. I believe there were six cats resident on that particular day. Seventeen-year-old Figaro suddenly jumped up into an empty chair at the dining table we had chosen as our interview site. Mary Jean settled him on the chair with a soothing pat. For the rest of the interview he sat upright on the chair, staring at me, then looking from me to her.

"He doesn't want your lap, especially?" I asked.

"No, he just likes to be part of what's happening." The purring was so loud we had to raise our voices. "He's got a world-class purr, that's for sure. It took me a long time before I figured that out, that he was just constantly circling around, not wanting anything but to be part of things. And when he's part of things, he does that purring. He doesn't go up on the table when I have people here eating. He just wants to be able to see."

Mary Jean noted that in Chicago, as in Northridge, cats are employed as mousers, especially in breweries, because breweries require vast amounts of grain, and vast amounts of grain still attract great numbers of mice. "There's an expanding industry for cats that live in warehouses and breweries and take care of the mice. In fact, we got an inquiry this week from someone who asked, 'Do you have an antisocial cat for our warehouse? We would feed and give water to this cat, this cat would have a comfortable place to sleep, we have no interest in or intention of interacting with this cat, we just want the cat for mice.'" So this, then, is the lot of the domestic cat and her feral relatives, to be both valued and despised for the same behavior, hunting.

I keep wanting to say that Mary Jean started Hyde Park Cats, an organization that funds programs to feed and offer TNR to feral cats, and to foster and adopt out cats in need of a home. But she keeps correcting me. She is not the founder, and the organization has nonprofit status, but it's not that organized. She is the treasurer and a member of the board, but she knows little about who is doing what on a day-to-day basis. She knows what she does—which, beyond keeping the books, is to go out every morning before work to feed the feral cats in two alleys, far flung from each other. She goes to the same spot, leaves some kibble on the ground, opens some cans and turns them upside down, pets a few cats, assesses the conditions of cats and their environs, then goes to work. "Covered in cat hair?" I ask, and we laugh, except for Figaro, who continues to purr at his place at the table.

Mary Jean did not set out to save cats, not even to feed them, much less to get involved in TNR. Her motive was to move more deeply into the human life of her own neighborhood, and the cats led her on a kind of magical mystery tour that paralleled what she describes as a rather proscribed middle-class commuter life. She met people she would never have met, and she developed relationships she would never have made, for better and for worse. And she became very involved with the cats themselves, providing transport and medical care, fostering cats who were waiting for homes. Though she may have started out in search of the humans in her neighborhood, she stayed even more for the cats.

"It's kind of a convoluted story. Back in 2008 I was working at the University of Chicago Divinity School and my colleague Terren called me up one day from her office on the third floor, and she said, 'Hey, I just saw on Marketplace there's this woman who wants to hire people, wants to hire someone to feed feral cats.' And I said these exact words: 'I have zero interest in feeding feral cats, but I'm going to call this woman up because I want to know what her story is.' I was curious, because I was only barely aware of feral cats at that point, so I wasn't into the story, I wasn't into the 'cats' narrative. So I called her up and we started talking and Jane had this story about these cats that she was feeding down the block from where she lived. Now here, I would ask you to not reveal the actual locations, you can make it up, say whatever you want. We try to be protective of the location because there are crazy people." As I listen to the note of secrecy, I think of Sabina's one-time cat feeders at CSUN, prior to the organization of Cat People, and of their desire to protect the cats they were feeding.

Mary Jean explained that Jane had been feeding these cats through a very difficult time in her own life. "She had lost her job and was having a hard time even feeding herself, but she had somehow managed to keep feeding these cats. Now things had started to pick up for her and she was planning to move away, but she just couldn't bear the thought of leaving without finding somebody to take care of the cats. And she was really concerned that they needed to be neutered and spayed and she didn't have any money and just was looking for help. She told me the story of her life. Her mother was killed and she was seriously injured, and her, I forget now, was it her mother, her mother's sister, her cousin . . . they were traveling to another city to visit family and there was a train wreck and her mother and her cousin, I believe, died. Anyway, she had a really tumultuous. . . ."

"So that's probably partly why the depth of her connection to the animals," I said. We both nod vigorously and say simultaneously, "Absolutely."

"She was self-conscious about that. She said, 'My family was just so messed up and my connection . . . I just feel like these cats are my family.' And I was deeply moved by the story. We bought a trap and we started trapping these cats, Jane's cats, and it started from there."

My interview with Mary Jean meandered into various alleys, meeting the neighbors along the way, one action leading to another, one relationship to another, all with a feeling quite different from Sabina's focused, chiseled campus program. Perhaps the differences in these two cat assistance programs has to do with the physical and social geographies of the places where the cats and their people live, as well as with the personalities of the people initiating the programs.

"Eight years later, and you're still doing it. Has the family grown?" I ask Mary Jean.

"No. In fact, in that location where we spayed and neutered, there was a

huge influx of cats there when we started feeding, and we spayed and neutered maybe fifteen visitor cats. The ones she was specifically feeding were a
family of a mother, her first litter which was two females and then the next
litter which was three males. So those were the five she was particularly
committed to. She had this dumpster that she had put on its side, where
the cats could go inside and get shelter. But then, once we started trapping,
all these other cats started wandering into our trap and as I started feeding,
then more of these cats started coming, but over the years there's been attrition. Some were there for the trapping and never seen again. When we released them they just went on their way, and then some died."

Mary Jean told me that it became clear to her, early on in the project,
that this was not only about feral cats but also about abandoned cats who
wanted to live in a human home. "From the very beginning, we had to start
making decisions about what we were going to do with these friendly cats.
One of them I just decided to bring home with me was my Mimi. She was
the matriarch of that little clan of Jane's cats. I had all five of the kittens in
here when they were being spayed and neutered and I really debated about
keeping them and trying to socialize them and I made the decision not to.
I've had lots of feral cats that were crazy wild, and they weren't like that,
but they were really very uneasy. I made the decision that I was going to release and take care of them. Now, eight years later, they are quite acclimated
to me. They come, I can pet them, they rub along my legs and sometimes I
wonder if I made the right decision. But now, to place them I would have to
place them together because they are so powerfully bonded. So I struggle
with that every day, especially in the winter."

Once again I hear the complex uncertainty of people who intervene in the
lives of nonhuman animals in order to help them. Our conversation has cir-
cumambulated, walked around in circles, wherein the tail of the cat meets
the tale of the human, like a circle dance—or perhaps like a spiral dance,
with both the cats and the humans making their rounds, maybe reaching
higher and better ground with every turn of the wheel, maybe just collectively staying alive.

"And cats," Mary Jean says, "I love the way they are in the world, they're
very deep and I love their sense of self-possession. I get frustrated when
people say that cats are aloof. Well, in some ways that's what I love about
them because I feel like, when I see so many dogs who are so emotionally
needy . . . now, I say this with hesitation because I've learned, I know my cats
well enough to be able to understand and see their emotional neediness too,
and it's intense, but it's not quite as much out on the table as with dogs."

I ask about some of the other volunteers.

"A thing that has really struck me as this work continued was that over the years we would get calls from various people in the community, people that I would never ever have the opportunity to get involved with because, let's face it, we talk to people who are like ourselves as we live our lives."

"So it's like an accidental alternative community?"

"Yes, definitely." She tells the story of Rosa, a Spanish-speaking woman who loves the cats but can't afford to neuter them, and who can't easily voice her inquiries to find out who might be able to help her with the cats. And of Carl, the elderly gentleman who sings in the church choir and leaves little notes poked into the fence near the cats. "Mary, this is Carl, the cat man, could you possibly spare me some food?" Carl now receives a delivery of cat food twice a month. "So this is what I mean, it's not just the cats, it's the people. So in some ways this cat work has jolted me out of the daily rhythm of my life, which gets to be so dictatorial, and you realize that you only see certain people, you only go on a certain path." There are some twenty people now, Mary Jean thinks, taking part in the lives of the feral cats, a handful doing the feeding and the TNR program organized and paid for by Hyde Park Cats. But most of the volunteers foster friendly cats who are waiting for a human home. There are students at the Divinity School who are involved, and local residents like Carl. For the most part, they don't know each other.

Sabina Magliocco's CSUN Cat People, first chartered as a student group, had been successful in its first years of operation, but it needed to become part of the life of the university in an ongoing way, because of course students move on and, if left to attrition, it would be a short time before a large cat colony reappeared. After Jessica graduated, Sabina wanted to create a vision for a program that would be sustainable and that would serve multiple purposes. The program had already had educational effects, changing the campus ethos about trash on the ground, relations with the neighborhoods, and animals on campus.

Sabina described to me a scenario with which I am familiar, that CSUN students are often the first in their families to attend college, that most of them work full time and also are full-time students. Many of them are parents of young children. Asking students who carry these burdens to come in to trap and to feed cats is just not realistic. So, after initial success in trapping and neutering cats, cat-feeding duties had begun to default to staff members again, some of whom had retired and returned to campus just for the cats. Sabina considered that something more was needed for the program to become an integrated part of campus life.

Of this transition time, Sabina reflected, "I knew I had to think of something else. When I taught my human-animal seminar, I wanted it to be a service-learning course, but it's challenging to find service learning that students can easily do. So we created a learning partnership with CSUN Cat People. We involved students from any service-learning class. One of the things that we did was a campus food drive. We decorated boxes and delivered them to various locations around campus. We would collect enough food to feed the cats for about half of the year. The rest of the donations came from pet food companies who gave us recently expired food that they can't sell."

There were parameters that made the service-learning partnership work: it was a semester-long commitment and therefore a limited number of weeks rather than an open-ended commitment. The fact that students were being evaluated and getting credit created a kind of support structure. It worked out that the service-learning project was enough to get sufficient student commitment, keeping the program alive and the number of cats limited to about twenty good mousers, who themselves were now integrated members of the campus community.

It was not only the cats who benefited from the program, but the students and, by ripple effects, the whole campus community. As students engaged with the cats, they learned to question old ideas and attitudes about their relationship to companion animals. One of the ways that this message infiltrated the campus culture was through educational presentations about the importance of spaying and neutering. Sabina explains, "It occurred to me that humane education needed to be part of a university education, in the same way that we want students to know about domestic violence or safe sex—to make students better-informed citizens. Humane education had to be part of that. We would put out an open call at the beginning of a semester to professors asking if they wanted to have a short presentation. The students would present, saying, "You know what? It's irresponsible to bring these pets to campus and then abandon them."

"When you were describing the students giving the presentations," I interjected, "I thought that, whatever effects their presentations had, they themselves were acquiring a deep understanding that would almost automatically cause them to become ambassadors."

"Oh yes! One student wrote in his final evaluation that he had come into the course resistant to neutering his dog, because he was afraid of taking away the dog's manhood, but then he came to understand things in a more complex way and had his dog neutered."

Students were often shocked to learn how many animals are euthanized at shelters, and a tide of concern for the animals themselves was swelling. "A new animal-control director came to Los Angeles County who completely revolutionized animal-care services by making a policy that the county would be a no-kill operation by 2020. And, by the way, this doesn't mean that no animal dies. Animals are still euthanized for a variety of reasons: they are too sick or wounded to save, they have behavioral problems like aggression, and so on. But the push is to work with local rescue groups, to let rescue groups take and foster animals for adoption." Prior to 2004, feral cats brought into shelters in Los Angeles were euthanized. In 2004, the City of Los Angeles began to offer discount vouchers toward the cost of spay-neuter for feral cats, partnering with nonprofits, and thus endorsing the idea of TNR.[22].Cat People was able to benefit from this initiative.

This turn of our conversation in the direction of community partnerships led me to ask if Sabina had received requests for Cat People to offer TNR in other locations.

"Yes, of course there were requests, but it's in our charter that we can't trap off campus. However, I often had people contact me and ask who can help. Alley Cat Allies, Stray Cat Alliance, and Best Friends: we strongly encouraged people to use these resources because TNR is effective."

Sabina suddenly sat up straighter and declared, "Now I need to talk to you about TNR in Los Angeles. It is still illegal in L.A. County. The reason it is illegal is because a group of environmentalists lobbied city council that TNR is bad for songbirds—which, I'm sorry, is a crock of shit. I fully understand and accept that outdoor cats, predator cats, hunt and have a negative impact on populations of songbirds, and also on populations of small reptiles and amphibians. But TNR is a better solution than euthanizing cats."

The reference is to a 2013 study that brought the impact of domestic cats on wildlife into public discussion with its estimate that somewhere between 1.4 and 3.7 billion birds are killed by free-roaming domestic cats in the United States every year. Add to that the figure of somewhere between seven and twenty billion small mammals. The authors of the study concluded that feral cats were "the single greatest source of anthropogenic mortality for US birds and mammals."[23]

"I've read about this," I interject, "but it seems to me that lack of TNR means there would be more hungry cats hunting, so I've never really understood the opposition of environmentalists to TNR."

"Exactly. Reducing the feral cat population and feeding the cats effectively reduces the hunting of feral cat populations."

Though the urban animals, we assume, may not know any more than do the whales and cheetahs about human legal jurisdictions, the humans who want to help them must understand and work within the context of those jurisdictions. "Luckily, the CSUN campus belongs to the state and not the county," Sabina says. "So state laws apply there. And there is no California law banning TNR, so we were completely within our rights, but other organizations around our campus were affected by the ban on TNR. And I want to say that the biggest negative impact on small-animal populations is habitat destruction, not the presence of feral cats. It's more complex than just the cats. On campus, the administration destroyed half the botanical gardens for a building project, and that was a huge habitat loss for birds, reptiles, and amphibians. You just can't put all the responsibility on the cats, there are other factors that have effects on animal populations, and human development of building sites is at the top of the list."

The Royal Society for the Protection of Birds (Britain's RSPB) interprets the findings about cats hunting songbirds differently than did the authors of the 2013 study. While acknowledging that cats hunt birds, they suggest that many of these birds are the weak, and the implication is that this is nature's way. Evolution provides that predators kill weak animals, leaving the strong to reproduce. "This may be surprising, but many millions of birds die naturally every year, mainly through starvation, disease or other forms of predation. There is evidence that cats tend to take weak or sickly birds."[24] Interpreting the extinction of species in terms of the hunting behavior of cats, the RSPB's conclusions mirror those of Professor Sabina Magliocco that, indeed, habitat loss is the single biggest factor contributing to species declines: "Those bird species which have undergone the most serious population declines in the UK (such as skylarks, tree sparrows and corn buntings) rarely encounter cats, so cats cannot be causing their declines. Research shows that these declines are usually caused by habitat change or loss, particularly on farmland."[25] Yes, cats are a problem for songbirds, but the songbirds and the cats have a bigger problem, and that bigger problem is human "development" of ever increasing measures of wild lands.

Beyond the kinds of problems that Sabina was able to foresee and forestall—the politics of intergroup relations on campus, the ways in which the campus program could and could not form cooperative relations within the larger communities of neighborhood, city, and county—there were other problems, fortunately few, of the human kind. Sabina impressed me as being more grounded and perhaps more practical than some animal-rescue workers. "As you know, I've also worked in wildlife animal care, and I believe that ambassador animals who can't be released should be sterilized because there is no reason for them to breed in captivity. These animals sometimes have injuries or behavioral problems that would make them less than ideal parents, and the last thing you want to do is raise another generation in captivity. Pets and any nonreleasable animals, I think, need to be sterilized. Of course, when you are talking about endangered animals, it's a different thing. But small house cats can dominate other populations, so they have to be neutered."

I wonder again my perennial question, can I know what it is like to be a cat, domestic or feral? Of course, cats and I share the basic body plan: a head, four limbs, respiration, circulation, digestion, reproduction. And we share the fundamental affects: seeking/curiosity, care, anger and fear, play, lust, and grief. But in many ways, I think I cannot know what it's like to be a cat because cats live in the strange liminal zone between wild and domesticated and have the capacity to maintain themselves across that spectrum. The capacity to take or leave social relationships is not something I have. The need for meat and the obsession with hunting that are so central for them are not in any way interests of mine. The differences between them and me are real; they can purr a self-healing that I cannot.

Though there are differences between a cat's fundamental motivations and capacities and mine, there are developmental events that we share across species. I think of my mother's elderly feral tomcats, Sassy and Fred (now minus some boy parts), for whom there has come a time when the work of life is less compelling. They no longer fight or hunt. While they haven't moved in, the old boys have slowed down to indulge in the comfort of their warm cat houses and soft beds under a covered porch. An older animal naturally seeks comfort and rest and the simple pleasures of the moment more than he seeks excitement and experiences of mastery. Feral cats, given the invitation, sometimes choose to retire from the rigors of living wild.

Sabina told me the story of the death of her beloved indoor cat, Idgie, and the opening it seemed to have created for Olive, the feral cat who had lived in her backyard and on her roof for fifteen years, resisting all invitations to become even a little domesticated. Olive's initial human connection started at the bowl of kibble and ended at the bowl of kibble, with just enough time in a crate for her to be neutered. "In February of 2014, we lost Idgie at the age of eighteen, after three years of congestive heart failure. Idgie was a cat of my heart, and I was inconsolable. We continued to feed Olive, of course, and one day about two weeks after Idgie's death, on a cold and rainy evening, I once again asked her if she wanted to come indoors. But this time, to my utter amazement, she said yes. I put her dish down just inside the back door, and she gingerly came in and began to eat. It was as if she understood that Idgie was no longer there, but more than that, it was as if Idgie had left her precise instructions on how to be a house cat. She immediately started using the litter box like a champ, never having an accident. She understood that she should not scratch the antique furniture or the Turkish carpets, but confined herself to a cardboard scratcher and some carpet scraps we put out for her. A few days after coming in, she discovered the fleece blanket on top of the down comforter on our bed." Olive never needed to venture out again, and she spent a glad old age with warmth and "poffiness," in Sabina's whimsical telling. "I will never forget my fairy cat," Sabina added after recounting Olive's death in her arms following a short illness. She noted that it was Olive who gave her the term *fairy cat* to replace the less pleasing term *feral cat*, and I have since adopted the expression myself because the whimsy fits. "Through her, I learned about and became involved in TNR. She taught me that even cats who have been feral their entire lives can make a choice to be tamed. I now question much of the received wisdom in TNR colony management which says that feral cats cannot become house pets unless they are taken before the age of seven weeks. And she loved me so hard, more than any other cat I've had."

It is indeed definitive that cats and dogs each have chosen humans as their companion animals, and that we've chosen them in return. One thing that seems abundantly clear to me is that we need to continue to do the research that provides us with factual and useful information so that we *can* care for them competently. Another thing that seems equally clear is that we *should* care for them as members of our extended family. Elizabeth Marshall Thomas, recounting a story about finding herself lying on the ground under the face of a feeding puma, and being surprised that she would ever put herself in such a situation, tells of a prior visit with this puma when the cat had taken comfort in suckling her arm. In that moment, the two beings, woman and big cat, formed lasting bonds of trust and affection. "Thanks to the soothing, the bliss, that we had experienced earlier, we seemed to understand each other. We had crossed our species' boundaries and had found the common center in each other, where all creatures rest."[26]

On the night before Sabina's departure from our home in Italy, a small group of friends sat under the veranda rocking contentedly with the children bundled in the laps of the adults. While Sabina played the guitar and sang, I saw from under her chair a black kitty paw poke up, taking swipes at the edges of her sweater where it overhung the chair. It was little Centaura, willing now to interact with Sabina, if surreptitiously. Perhaps it was the purr of the music that engaged her. Sabina sang an Italian folksong about a young soldier who, walking among the red poppies on a spring day, encountered an enemy soldier on the road, and paused, stunned by the realization that the young man he met was so much like him. In that pause he was shot and lay bleeding to death among the red poppies. And I will always wonder if Centaura might now be climbing her olive trees, pouncing among the red poppies, and eating her mice and lizards . . . if she'd not stopped to love me.

CHAPTER 8

Love Dogs

As a child I often sat mesmerized by a framed print on the wall at my grand-father's house, the image of a little boy clutching a springer spaniel to him-self, tears running down his radiant face, his knuckles white with fierce clinging. A love that was lost had become grief, and then the love was found again. The image of that sorrow and of that healing filled my psyche. It was so compelling that, twenty-five years later when I learned of a litter of springer spaniels for sale from a breeder two hundred miles away, I could not wait to go get the last of those puppies, Eleanor Roosevelt. Now I under-stand that the reunion of the boy in the painting and his dog revealed to me my own vocation, that I would have to hold that image, absorbing it, for five decades until I would come to completely understand it.

I have learned over the years that I am not alone, that there is a dog in the center of many people, planted there like a seed, secretly connecting us humans to the other animals and secretly feeding our deepest hunger for animal belonging, a belonging to and with a bigger world of relations. Jane Goodall famously credits her childhood dog Rusty with planting the earliest seeds of her renowned vocation as scientist at Gombe and later as world ad-vocate for animal conservation and rescue. I believe this sense of belonging in the family of animals is qualitatively distinct from human belonging, that it can provide for an immense resilience, and that many people first experi-ence that belonging in the company of a dog.

Not all psychologists agree on the genuineness and value of affiliation across species. John Cacioppo and William Patrick claim that affiliation with dogs and cats is a stopgap measure for lonely people, a kind of second best to human company. Notice what seems to me the strange logic in their reasoning: "One of the lessons of Hurricane Katrina was that pet owners were so committed that many were willing to risk their lives to remain in the city to care for their animals. Was it the sense of being left alone with the elements—in a sense, rejected by those who left the storm—that made their attachment so strong? . . . Perhaps many of these economically deprived people had felt rejected all along."[1] In the first sentence the courageous humans are so committed to their animals that they won't leave without them. By the second sentence our heroic humans have become victims of human abandonment themselves, based on no evidence and in contradiction to the fact that they themselves refused to leave their animal companions. And by the final sentence, these heroic humans are diminished to the realms of the very other, poor people who felt rejected all along. Poor dears. But people who value the company of animals say otherwise. In fact, research increasingly shows that many people prefer the company of other animals to humans. For me, the pertinent question to ask is why this is so. And my answers to the why question lead in the direction of belonging to a world beyond the human and to a deep and old source of resilience that lives in that extended family.

I think it is no wonder that dogs received the moniker man's best friend, because dogs do love us; their loyalty is legendary. Their love for us has also been repeatedly scientifically investigated and affirmed. The oxytocin level in a dog's blood goes up almost 60 percent when they see their human, and that physical fact represents the truth that many people intuitively know, that their dog loves them. The reward centers in the dogs' brains light up when they smell their human.[2] In other words, the same biological mechanisms of love and bonding operate in both dogs and humans. In dogs, though, the feature of interspecies love is so distinctive as to be considered the specialization of their ecological niche, says psychologist and behavioral scientist Clive Wynne, who has spent his career investigating the social nature of dogs.[3] "Dogs fall in love much more easily than people do. . . . It's not the case that dogs have special genes or special capacities to form relationship with humans. Dogs just have special capacities to form relationships with anything. Whatever they meet early on in life, they will then accept members of that species as potential friends later on."[4]

Dogs loved me first, and I never questioned loving them back. I think that is part of the strategy of the species *Canis familiaris*. Granted, our two species share some behavioral traits that make for a good partnership. Wolves

and humans pack-hunt during the day, and both species maintain emotionally bonded social hierarchies. Dogs evolved from wolves through a process
called neoteny, expressing the traits of their wolfy childhood in adulthood,
traits like curled tails, floppy ears, and docile dispositions. The babified
wolves that we call dogs allowed humans to become their "top dogs." Cleverly, they assigned to us the adulting work. Research in psychology has
shown that when adults see baby faces and baby traits, they respond with
the feelings and behaviors of care.[5] Dogs might be said to have exploited this
feature in humans, with the consequence that the two species have lived
together in emotionally bonded relationships for at least fifteen thousand
years. There is evidence of ritualized human-dog relationship going back
125,000 years. The bonds of care between humans and dogs go deep, so deep
that we may even be seen as coevolved. The thirteenth-century poet Rumi
was speaking for the ages when he said, "There are love dogs no one knows
the names of. Give your life to be one of them."[6]

The Chauvet Cave in southern France contains the footprints of a human child and a wolf-dog walking side by side thirty-four thousand years
ago, and in the same cave a pair of wolf skulls were placed on either side of
the cave entrance. Yes, most of us have heard that dogs evolved from wolves
and have a lot of genetic diversity, but consider this, that dogs must have
had a role in their own domestication. "An evolutionary process resulting
in the domestication that was not initiated directly by humans rewrites human history."[7] Human history, in other words, has partly been written by our
companion species, agents on their own behalf who chose to partner with
us. It is fair to say that dogs domesticated us as much as we domesticated
them. At the level of species our relationship with dogs might better be described as a partnership of mutual origin and mutual benefit.

Dogs belong by evolutionary path in the close company of humans. They
were the first domesticated animals, long before the agricultural revolution.
And, in contrast to cheetahs, whose lack of genetic diversity imperils their
breeding capacity, dogs are genetically extremely diverse. On at least one
variable they are the most varied species on the planet: adult dogs range in
size from under three pounds to two hundred and fifty pounds.[8] With such
genetic diversity dogs have the capacity to specialize in a variety of useful
roles, from guardians to hunting partners to emotional-support animals. Allowed to breed naturally, this genetic diversity also allows for endlessly entertaining dog forms to be expressed in bodies that sometimes appear to be
put together from spare parts.

Sabina Magliocco tells me that dogs have this creepy and unnerving habit of asking her to tell them how they should feel. She doesn't want that impossible level of responsibility! "Shouldn't they know how they feel?" I tell her she misunderstands, that dogs want to know how *you* feel, they want to know you're okay, because if you're okay, they can go do what they need to do to be okay. This need to assess the emotional okayness of their social worlds is what the infamous dog butt-sniffing is about. When dogs stop in the middle of a social interaction to do a butt check, sniffing round the rear and sometimes also the chest of the other dog, they are taking the emotional temperature, assessing the mood, and adjusting their behavior accordingly. Dogs, I like to suggest, are metaphorically the perfect little wives and husbands, moms and dads, the comforters and protectors that most everyone longs for. They will cuddle you even when you are drunk, comfort you for being ridiculous, and risk themselves to protect you. They will stay by you when you are sick, in the crook of your knees or the small of your back. Dogs will run for help when you are in trouble. They will wait for you at the train station. They will keep vigil at your grave. They will even let you go somewhere else to do something without them, provided you return within fifteen minutes: longer, and you might find some anxious widdle on the floor or a thoroughly chewed couch cushion. With such legendary loyalty and devotion, it's no wonder that dogs sit at the center of so many human hearts. Another and perhaps equally valid insight is that dogs seem to have written the playbook for what is often called codependence, the relational style that facilitates bad behavior in others, even at one's own expense.

I used to say to my Jack Russell terrier Bronwyn, "It's a good thing you have a psychologist for a mommy, because you really need one." Would that being a psychologist rescue mother had been enough to save her. I will never forget taking her to our well-loved country vet, Dr. Vicky, waiting in the waiting room with all the blue glass stuff on the shelves, with Christian rock wafting from a radio and a fresh scripture quote written on a whiteboard each day. Many people were at Vicky's because this ambience caused them to trust her. I was there in spite of it, having learned that she was an amazing human being and an excellent vet, one whom the animals trusted. I took the trust of the animals as a message not to pigeonhole her. She treated the animals with kindness and dignity, communicating their problems to their humans in terms of empathy. "Lexi has a hard time swallowing, and she has had a tummy ache every day for a long time," she once told me in order to explain the subjective meaning of injury-induced esophagitis. Not surprisingly, the animals responded to Vicky with cooperation. Except for Bronwyn.

I had waited six months after my beloved Molly Brown died to look for another Jack Russell terrier because I did not want the new dog to be saddled with expectations that she be like Molly. When I decided that enough time had passed, I looked for Jack Russell terrier (JRT) rescues online. I filled out a form that went out to a broad group of organizations, indicating I wanted a JRT female puppy. Haha, came the reply, good luck with that one. You want a full-breed JRT female puppy, without baggage, right? We will keep your request on file. That was my first clue and, in my eagerness, I missed it. Two days later, I got word of a fifteen-week-old puppy available for adoption, if I passed the test. With visions of Molly Brown dancing in my head (in spite of myself), I gladly signed up to drive three hundred miles to take the test and pay the price to bring that puppy home.

When I arrived at the farm in northern California where the puppy was being fostered, Bronwyn's human foster mom brought her out to a picnic table to meet me. She set her on the ground. We talked as I petted the puppy, and she noted that this dog was nervous around children and should not go to a home with children in it. Finally she suggested I pick up the puppy. I did so gently and the puppy licked my hand while the foster mother sighed with relief, and offered me papers to sign. I signed. Then I packed up the little dog and her vaccination and spay records, and off we went into a new life together.

I quickly learned that we would be less together than I'd hoped. Bronwyn wanted to be alone. It was very difficult to bond with her because she avoided contact, sleeping under the clothes in the back of the closet and snarling when approached unawares. Food was her only interest, and she went into situational insanity in proximity to food. It reminded me of the opening scenes of the movie *The Miracle Worker*, in which Patty Duke plays the young Helen Keller, running wildly around a dining table, stuffing food into her mouth while Anne Bancroft plays Annie Sullivan, assessing the work before her and deciding to do it. With that scene and Annie Sullivan as my inspiration, I decided to work with Bronwyn. I experimented by removing all food, putting a piece of bread in my mouth, and directing her to come and take it. I was terribly relieved when she did not snarl, much less snap, but came rather timidly to snatch the bread and run away with it. I began to pair this exercise with pointing to my eyes until she made eye contact. And so our love for each other was born of food, eye contact, and goodwill, actually a pretty good recipe for love in general.

But that cleverly constructed love was not enough to help Bronwyn pass her puppy test. When Dr. Vicky moved her new patient on the examination table, Bronwyn responded by snarling and biting me. Vicky looked up startled, took a step back, and addressed me pointedly. "Anne," she said, "you have to take this dog back. This dog needs to be on a farm where she can live outside for exercise and have lots of space without too much stimulation. This dog can't take any more, she's full up, she has anger issues." I still have the record that reads, "Failed puppy test: anger issues."

My heart sank as I walked back to my house with Bronwyn in my arms. If I took her back to rescue, I felt sure she would be euthanized. I decided to keep her, to work with her. I knew from her foster mom and from her papers that she was born on a farm in Kentucky, that her breeder/owner had accepted a free airline trip in exchange for the little dog. The airline employee brought the puppy home to California for his three kids and within three days was looking for a foster home for her. Leaving Vicky's with Bronwyn in my bloodied hands, I understood that the puppy had bitten a child. That was why she'd been available for adoption.

Bronwyn had anger issues that, over months and years, I came to understand probably derived from two conditions: complete lack of socialization in puppyhood and physical pain to such an extent that anger was deeply programmed into her brain during her infant and childhood development. When she simply quit walking at about two years old, I got her to a spinal orthopedist, who concluded his analysis of various images of her spine with the opinion that she had either been thrown or had something very heavy drop on her during her infancy. I understood then that she had likely been separated from her mother and siblings because of her injuries. The other puppies matured and were sold. Bronwyn was kept somewhere alone, with high levels of substance P flowing through her immune and nervous systems. Substance P is a neurotransmitter that communicates pain information within the central nervous systems of all mammals. It also interacts with inflammatory cells of the immune system, and it activates the rage system deep in the brains of animals.[9] Bronwyn, like many people in chronic pain, was habitually angry.

The orthopedist showed me the images that told him she had eggshell-thin bones, just enough to work with, and he rebuilt her spine. After surgery, to her dismay, she was kept immobile for weeks before she walked freely again. Then I began to teach her some complex tricks, which she enjoyed immensely: picking up something I tossed and returning to drop it at my feet, walking in a circle around me, and coming to sit down next to me, the two of us facing the same direction. Bronwyn was almost normal for a few months. But Bronwyn never got over her anger issues and, like some humans, that was because her anger was an outgrowth of physical pain that had occurred in the developmentally formative time of her life, and in social isolation, a recipe for recurrence throughout her life.

All of this is to say that Dr. Vicky was right about Bronwyn, but the fact that she was right did not make me able to abandon the little dog. No, I would be there with her and for her for as long as I could be. Unwittingly, I would facilitate her bad behavior at my own expense. Every day for five years, my first words on returning home were, "How is Bronwyn?" Anxious mom to avoidant dog. Until one day, when I saw that Bronwyn's now terminal pain had become too much for her exhausted little body to bear. I had noticed her rapidly decreasing mobility, her increased edginess. I knew that it was time for her to leave us, that she was on her way out.

I'd spent the night in the guest room with Bronwyn in a crate next to me, snarling, after having snapped in pain at me the previous evening. I said goodbye to her as tenderly and calmly as I possibly could, with all the complex emotions of such a moment swirling inside me. I don't believe in reincarnation, but with utmost tenderness and sincerity I invited her to come back to us in another life. Then I let her go. At long last, I had had to surrender Bronwyn. I had had to give up: there was nothing more I or anyone else could do for her.

Bronwyn's anger issue is probably better labeled as rage because it was very primal, not based on ideas or other emotions, not rooted in a sense of unfairness or jealousy, for example. Bronwyn's rage wasn't about anything, it was simply a response that occurred at the level of her feeling her very existence threatened and instinctively acting to counteract that threat. The rage system in the brain is one of the seven primary affective systems described by Panksepp and Biven, those impulses so deep in the brain and so deeply rooted in the evolutionary past, so immediately connected to the body, that they can be called instincts because they come with both a felt sense of response and a physical disposition to act in certain programmed ways. In the case of rage, the compulsion is to fight, to dominate the perceived enemy for long enough to get away from him. This behavioral predisposition is expressed with clenched jaws, snarling, snapping, flexing of muscles, and "making big body." Of the rage system in the mammalian brain, Panksepp and Biven write that it "is not fundamentally designed to punish but rather to bring others into line, rapidly, with one's implicit (evolutionary) desires."[10] Rage is about getting immediate control in a life-threatening situation.

Bronwyn had grown and developed as a child alone and in pain, probably limited in mobility for some time after her initial injury. Both physical pain and social isolation trigger the brain's rage system: recall that loss of social connection can even lead to death from grief. In fact, being connected to others is a condition of safety, and we are unsafe without that connection. For most of us mammals, the need to be connected to others remains as real and deep in adulthood as the need for food and safety and, because of the depth and reality of the need for connection, rage and grief can each be triggered by the lack of connection. In fact, endogenous opioids and oxytocin are the two natural substances that quiet rage. Endogenous opioids, the feel-good chemicals produced in the brain, provide a direct biological reward, they are the neuromodulators that make us feel good and that allow us to become addicted to substances that mimic them, such as heroin, or the synthetic opiates used to treat chronic pain in humans. Oxytocin is the neuromodulator that makes us feel good socially, connected and trusting. Both the trust hormone and the pleasure hormone calm rage.

The neural system for rage is rooted deep in the oldest parts of the brain, in an area called the periaqueductal gray, from where it routes through the medial hypothalamus, the brain region that modulates physical homeostasis, and from there to the amygdala.[11] The rage center in the hypothalamus was first discovered by Walter Hess, whose experiments located the rage system in cats by stimulating the cats' hypothalami with electrodes (work for which he received the 1949 Nobel Prize in Physiology or Medicine).

As an important aside, based on his own experience in conducting these experiments, Hess believed that the cats experienced both the feeling of rage, the emotion, and the bodily actions of rage: increased blood pressure, muscle tightening, and readiness to bite or tear, kick or claw. However, in his reporting of the research, he denied his real perception that the cats definitely had internal experience of the feeling of rage, and he publicly conjectured instead that the cats produced a mechanical reaction, not a felt response. He effectively lied about his own Nobel Prize–winning scientific investigations because he did not want his work to be summarily dismissed by the behaviorists who dominated biopsychology at the time.[12] I point this out to underscore that the culture of science—the fundamental assumptions, the shared rituals, historic metaphors and memes, and the usual run of social pressures—influences the results of science, and has especially had a long and inhibiting effect on our study of other animals. Fortunately, science is a method that is self-correcting, and it has largely corrected itself from wrongly held fundamental assumptions based in the social psychology of human beings, like the downward comparison to other animals that makes us feel better about being humans, the assumption of human distinction and superiority.

The hard facts of neuroscience have turned the table on that model, re-placing conjectures about what the behaviorists called "the ghost in the ma-chine" to describe an animal's internal felt sense of life. In other words, ear-lier scientists essentially refused to consider an internal felt sense of life in other animals, until such consideration was unavoidable because scientists came to realize they had held what psychologists call a "premature cognitive commitment," or an unsubstantiated assumption.[13] Today we know that the internal and subjective felt sense of life—the experiences to which philoso-phers refer collectively with the term qualia—is a primary aspect of all ani-mal life. All animals have a sense of the quality of experiences of living. At the most basic levels these experiences are pleasure, toward which we animals move, and pain, from which we animals move away. Even bees and cock-roaches have shown this level of affective preference and avoidance.

Not only can we assume that all animals have an inside, a felt sense of being in their worlds, but they communicate about that experience to oth-ers, outside, such that our insides are activated and motivated by the world around us, including other animals. We can comprehend other animals' in-sides because we have shared features of brain and body across animal spe-cies and also, of course, because we can listen to the communications of others of our kind. But beneath language we have this ability to understand the experience of others, to empathize, that is based in shared biology. The biological foundations of empathy are wired into us and happen body to body, through mechanisms like mirror neurons and muscle mimicry.

Mirror neurons occur in the brains of many species and activate when an animal receives sensory input of action in another animal, replicating the motor activity of the animal observed in the brain of the observing an-imal.[14] You watch me type and your brain offers you the experience of my typing within your own experience; you subliminally feel your own fingers typing when you see me typing. In the case of muscle mimicry, it has been found that a great deal of information passes body to body, in the absence of words, because we subtly and unconsciously imitate people and animals we encounter. Explaining the underlying biological processes that create the platforms for morality in animal lives, Patricia Churchland summarizes, "Psychological studies on unconscious mimicry in humans show that pos-ture, mannerisms, voice contours, and words of one subject are unknow-ingly mimicked by the other. Such mimicry most people do regularly as part of normal social interactions."[15] In other words, you don't need to speak the same vocal language to have empathy with another, even across species lines. You just need to live in an animal body and pay attention to the world

around you. Or perhaps it is more accurate to say that your own body is listening to the bodies around you all the time, but you might be distracted
from the music of life by the lyrics running through your head.

Certainly Bronwyn had an internal and felt sense of life, and even of connection with me, but her internal states were dominated by a desperate pro-
pensity to rage responses, programmed into her in infancy and then main-
tained by a continuous flow of substance P. Bronwyn loved me and Bronwyn
knew I loved her, but that was not enough to save Bronwyn from a very
tough life and an early death. I have often asked myself if I did her any favor
by saving her into a life of pain and confusion. I simply do not know the answer to that question. I do know she had moments of joy and pleasure and
of loving her life, even loving us. Does that count as some kind of salvation?
I don't know. So, of course I have asked myself why I kept Bronwyn at all.
Why did I not heed Vicky's advice and take her back to JRT rescue? Very few
things about Bronwyn are clear, even these many years later, but this was
always clear: I tried to save Bronwyn because I owed a debt of gratitude to
dogs, and specifically to JRTs, because the dog who saved me, Molly Brown,
was a JRT. I was paying it forward, a sentiment I heard expressed repeatedly by people working in animal rescue. In the past, this fact was often dismissed as silly sentimentalism, but now we understand the ways that affects
and emotions create a mutually regulated web of relationships within and
among species.

I must be honest enough to say that I made mistakes with Bronwyn, the
greatest of these being to ignore my own clarity, based on listening to her. I
knew this dog needed to rebuild trust, carefully and slowly, with wide mar-
gins for error. Early on—soon after the eat-bread-from-my-mouth intervention that bonded us—I hired a dog trainer. Because Bronwyn was generally
not attentive or responsive, the dog trainer was amazed that, even uncon-
tained and at a distance, Bronwyn would sit when I told her to sit. That was
evidence of the trust we'd established. The dog trainer next told me that I
needed to establish myself as top dog, using techniques that I thought might
inhibit trust in my broken little dog. I did try those techniques, and I think
they did inhibit trust; they set our relationship back. Recall that the biological purpose of rage is to gain immediate control. How, then, might dominating the enraged animal not cause harm?

I should have trusted my own instincts to make trust-building primary.
Just as with human psychotherapy, there is no one-size-fits-all way to train
dogs, and rank within the pack is not the only factor in a dog's social life.
Some dogs need physical and psychological containment, others need the

freedom to explore, the space to retreat from contact, and only gentle tone and touch. Some need not to be touched at all. Bronwyn was an angry little dog in perpetual chronic pain who needed space and tenderness, both at her discretion, not mine. But even had I perfectly dog-mommed her, physical pain and the rage it facilitates would likely always have dominated her little life. Bronwyn was programmed for avoidant attachment. She needed space and to choose her pace, not to be smothered under someone else's idea of occupying the position of leader of the pack.

As for disorganized attachment, dog shelters are full of animals who exhibit this style—they don't know what to do to get it right with the humans who are both dangerous and beneficent.[16] It happens to cats, too, but cats, though attached to their humans, are less dependent on their relationships with humans.[17] Shelter and rescue dogs need to be patiently taught by someone who can listen to them, understand them, and consistently show them how to get it right. Sometimes that never happens: some animals are so anxious or so injured that they never develop the ability to attend and relate to even the kindest and most consistent of rescuers. But dogs seek to please humans, so it is often relatively easy, with consistent communication, to rehabilitate an anxious and crazy dog just by communicating what it is you want from them. This simple realization is what lured Laura Coulombe into the world of dog rescue.

I met Laura, my niece, for the first time as she was turning forty. Adopted as a newborn, she grew up with no knowledge of or connection to our family. When we finally were connected we found all manner of family resemblances, not the least of which was a love of animals and a determination to offer them care. Laura has a bodacious sense of humor, runs a national dog-rescue organization, and is also one of the tribe that never quits trying. I spoke with her three times over a period of three years about her work, learning in the first conversation how she came to found a dog-rescue nonprofit. In the second, I heard her describe feeling the burden and burnout that comes with the work. She was promising then to scale back, to establish better boundaries, to make room for other things in her life. On the occasion of our third conversation, she was back at it full tilt. Tossing her long auburn hair back over her shoulder, she tells me in conspiratorial tones, "What they don't tell you is that you can never leave." That sentiment is something I heard over and over again, that it is impossible to leave the work of care for other animals once you have engaged in it, that it is a burden as well as a joy, and one that cannot easily be set down.

Laura tells me that she herself has begun to write a book, a truthful book, about animal rescue, about the ways in which the work changes human relationships and even one's sense of being a citizen. She tells me she has been arrested twice, once for carrying puppies who were given health clearance in one state but had begun to express symptoms of parvovirus while being transported. Having crossed state lines, Laura brought the now sick puppies to a vet, was reported, and arrested. She speaks of missing important events in her human family's life because of the choice to attend to urgent rescue needs. She mentions the president of a sister rescue nonprofit, Barbara, who missed her own mother's funeral because of compelling dog needs. I told Laura that I thought her book needed to be written, that she has the frankness and humor to speak of the very complex human aspects of dog rescue. She replied that she is writing because she wants to get it on paper, to see it herself, to perhaps understand the work and herself in the process.

Laura grew up with dachshunds and with cats. She says that at about thirteen years old she began to identify as a cat person and had cats throughout her adult life. But in the context of a divorce, she moved across the country and bought her first house. Then she bought two Italian greyhounds at a pet store. When she brought them home to the Victorian house she had purchased to restore, she found herself unable to housetrain them. Finally, unwilling to surrender the house restoration project, Laura decided that she just could not keep the Italian greyhounds, and she sought out a rescue for them.

The owner of the rescue asked pertinent questions about what training Laura had tried and what the consequences had been. In the process, Laura learned about belly bands and pee pads. She learned to analyze problem behaviors and, she says, she was hooked. She had become intrigued by the sense that there were creative ways to work with dogs so that the relationships worked for both the humans and the dogs, resulting in interspecies families staying together. The "Iggies" she had sought to be relieved of, Ella and Jack, stayed with Laura for the rest of their lives. The Victorian house never got renovated.

Laura went on to foster many dogs and to serve on the board of a rescue nonprofit, learning the ropes: the business models and legal requirements and health issues that are common in dog rescue. She encountered the conflicts and pitfalls of organizing and networking with other humans doing dog-rescue work until the day came when she thought she could improve on what she'd seen to date. Being of entrepreneurial and independent spirit, she decided to establish her own nonprofit dog rescue. For practical reasons of establishing herself in a niche, she chose dachshunds, the dogs she knew best from childhood. Early on, she accepted forty dogs in one single rescue from a breeder-gone-hoarder situation. The owners of all these dogs were elderly and no longer able to care for them adequately, and the dogs were in declining physical and mental health. "Forty dogs?" I ask. "What do you do with forty dogs? Where do you put them?"

"In the house," she responds matter-of-factly. Laura continuously exhibits this strange quality of being utterly pragmatic and doggedly heart-driven at the same time.

"Forty dogs?" I repeat. "Are the walls of the entire house lined with crates?"

"Basically, yes."

Furever Dachshund Rescue (FDR) is now a national rescue organization, with fosterers in northern California and New England. Although dedicated to rescue of and education about dachshunds, Laura, a self-described bleeding heart, takes other dogs when there is a need, and finds loving homes for them too. FDR is a volunteer operation with no employees. No one is paid to do the various works: going out to pick up animals for assessment, having any injuries treated, making preparations for adoption (inclusive of vaccinations and neutering), fostering animals who are waiting for adoption, transporting dogs to their Furever homes, sometimes in distant locations. Adoption fees and donations are the resources that pay for the veterinary care that the dogs require.

It all sounds grueling. Why does she do it? Laura's voice catches as she describes taking dogs away from kill shelters, and the feeling of despair in leaving so many dogs who take their final walk from the kennel to the euthanization station, who seem to know where they are going, to their deaths, because no one will have them. Ouch. Yes, we can feel that pretty directly, the profound grief of facing death simply because of being unwanted, a nightmare for us social animals. "The walk that these dogs take, one by one, all day, every day, it's wrong. And that's why it's so important to educate people about rescue, it's the chance to teach people who want to surrender their dogs. 'What seems to be the problem? Are you open to working on it? Can we help you solve the problem?'"

As I listen to her, I think: these are our evolutionary partners in existence, left to die because they let us assume the adulting. Laura says that one of the hardest things about doing dog rescue is the feeling of being contaminated by the filth of other humans. "It really disgusts me that we think we are so superior. Take away our guns and our cars and we don't stand a chance."

I can't help but ask Laura the obvious question, whether all this rescue work with the particular spin of keeping interspecies families intact doesn't somehow come back to her having been herself adopted as a baby. Is she working out something about that, I dare to ask her, understanding only too well the import of the question. "No," she replies with her customary confidence. "I had a normal family life, a mom and dad, and I grew up with everything I needed. I didn't think about it. No, it's not about that, at least not consciously. I have always been a person who likes to help. Before this I volunteered in a convalescent hospital. But with dog rescue, everything is last minute, urgent. You'd better plan to not have plans."

Laura's inclination to help has been centered on dogs, and dachshunds in particular. She continues to operate one of the newer styles of dog rescues: breed-based, with more than one location and with foster families scattered in far-flung places. She drives a truck with a trailer. She's the proud owner of a lot of dog crates, and toys, and food bins and bowls, and collars and leashes and, yes, belly bands and pee pads, and she's connected to a network of veterinary practices. She lists dogs on her own website and on Petfinder, an online database for animals searching for families. The idea of this model of dog rescue is to get people to swear off buying dogs from bad breeders, puppy mills, and their pet stores as sources for their companion animals, and to adopt instead some of the desperate dogs looking for homes. Laura has somewhere between ten and twelve dogs who are unadoptable for one reason or another: illness, old age, behavioral problems. These are her sanctuary dogs who will stay with her for the rest of their lives. She proudly tells me that she has never been bitten by a biter dog, that she learned how to be an alpha. "I ignore every dog I meet until they come beg for my attention." Her voice trails off with a hint of uncertainty. "Dogs I can do. I think I would be good just living with dogs. . . ."

Recently, Laura told me she has rescued a horse whose legs were intentionally injured, "sored" by use of whips, in order to make touching the ground painful so that the horse will pick up its feet faster and higher, all to create a "pretty," a high-stepping gait for the human gaze. Is there no end to human depravity, I wonder, learning of one more gratuitously torturous process invented by humans for trivial purposes. Laura is paying a vet to heal Joshua's sored legs, and paying a friend to board him, and she has bought a trailer to transport him. She who in our second conversation was swearing off the rescue business has just expanded her participation.

I also wanted to visit what I think of as the old-fashioned kind of shelter, one that is based in and serves a particular neighborhood or region and that is more general than the breed-specific and web-based rescues. As we drove north out of Bend, Oregon, my friend Betty told me about one of her favorite charities, Three Rivers Humane Society (TRHS). We meandered the country roads and Betty, a lively widow in her late eighties with a sharp sense of humor and stubbornly passive style, meandered through a story about how she came to be connected. That's Betty, always has been Betty, slow, meandering, going where she's going at her pace, not yours. She still speaks with a long, slow Arkansas accent, even though she left Arkansas some sixty years ago. She just winds me through the back roads, literal and metaphoric, until we reach our destination or forget where we were going and stop for ice cream.

I gleaned from the conversation one fact of particular interest to me, that this shelter recently received fifty-seven Australian shepherds from a breeder whose operation was broken up by local police for violation of animal welfare laws. After receiving warning citations, the breeder was arrested and charged with criminal animal neglect. Some of the dogs were pregnant. Each dog was assessed and treated, if necessary, by a local vet, and then the TRHS organized the transport of all fifty-seven dogs.

In order to accommodate the sudden influx, the shelter had had to scramble to reactivate enough of its outside kennels to accommodate double the usual number of dogs, some of them being treated for health and behavioral problems. The outside pens had to be refitted with siding and roofs to protect the dogs from the summer sun, with water misters to keep them cool, with blankets and doghouses to give them comfort and shelter. The paid staff of seven employees was working double time to serve a doubled population who needed to be fed, walked, taken to group play for the socialization that is so necessary to a dog's well-being and future placement, cleaned up after, comforted, assessed, and gotten to medical or training interventions every day.

We pulled in and parked next to a tidy enclosed trailer that had been made into a fundraising recycling wagon, complete with tasteful sponsorship advertisements, and filled waist-high with aluminum cans visible through the service window. We walked past the grassy playground, a kind of small dog park where the dogs came to frolic, a few at a time, and we entered the large reception room. There was a young volunteer of maybe ten years old, cheerfully breaking down cardboard boxes between bouts of rolling on the floor with a vivacious nipping and yipping puppy. The receptionist called for Jerilee Drynan, director of operations, who appeared in uniform of jeans, sneakers, rumpled logo T-shirt, dark circles under tired eyes.

I had not thought to question my notion that this was a local branch of the Humane Society, a national U.S. organization whose mission is animal protection. Jerilee quickly corrected my misunderstanding, explaining to me that local humane societies are not affiliated with the national animal advocacy organization, that each shelter is a private nonprofit. She offers that some shelters choose the name because it communicates what their organization does, namely to be decent humans, rescuing animals in trouble.

Three Rivers Humane Society is an independent local no-kill shelter that has a contract with Jefferson County to receive and care for animals brought to them by community residents and county agents, including all strays from the community, ten-day bite holds when a dog bite report is received, and animals who are in cars when police arrest a driver. Jerilee explains, "We don't go out and get the animals, but we receive them when community members or county officials bring them in." While we are talking in the reception area a staff member comes in the door, having just walked for two hours from the reservation where she lives to the shelter where she works. Podzi declares it was a rough night with her man, that she had cut short a human fight and turned to her dogs for comfort, then got up and walked in to do the work she loves.

I asked Jerilee how long she had been involved in animal-rescue work, and she proudly responded that she had opened a shelter, Alaska Humane Society, with her dad when she was still in her teens, then gone on to study child development and work in early education. That's the setting, in fact, where Betty and her partner of more than fifty years, Caroline, had met Jerilee. Caroline had served on the board of the local women and children's organization, and Betty had volunteered on projects to improve the children's school environs. When Jerilee returned to animal care, Caroline and Betty came with her. Jerilee had a vision to care for homeless companion animals and to be an educational force in her community, especially teaching children humane education about the importance of lifelong relationships with companion animals, and about the obligation to spay and neuter them.

Jerilee looks tired as she tells me that she focuses on dogs and cats "because they're the ones that need the most help, and I can't help everybody." Sure, she takes other animals when they are brought to her, but she adds, "I mean, I love horses but I can't do horse rescue, I can't do ferret rescue, I can't do bird rescue. This is a job that is very easily a situation where you can burn out and so you can't spread yourself so thin that you feel hopeless. It never ends. And there's this feeling you are always cleaning up other people's messes. Yesterday I had a guy say, 'Okay then, I'll just take this cat out and shoot it if you're not going to take it.' Some people love to say things like that. They bully and threaten. They don't recognize any limits, like a finite amount of space. I sent him away. And still, I let more people push than I should."

I ask if she and her husband, who run the shelter together, ever have the chance to get away. She bursts into laughter, then sputters to a "Nnnoooo." She explains that they get puppies with parvovirus, who have to be kept separate from the other dogs because parvo is so contagious and so deadly. Without treatment, 80 percent of puppies with parvo will die from it. Here they have treated more than four hundred puppies for parvo and only lost twelve of them. "But it means we have to be here every day to give them fluids and antibiotics, to clean them because the staff can't touch them. We go in there gowned up and following sterile practices. Some days we say we are not going to talk about work, but that lasts maybe ten minutes, because it's our life. It never stops. Yeah, I am tired. But you know what? I am gonna get up and do it again, and I am going to do that till it's done, which is forever."

Betty and Jerilee declare that it's time for the tour. They walk me around, showing me kennels, examination rooms, crates, washing machines for dishes and laundry. Jerilee says that underneath all of her work in life is education in kindness. She believes in kindness, she teaches it, and she won't work where it is not valued. She walks us outside to the newly refurbished kennels full of little Aussies. The puppies are darling, many of them having one brown eye and one startlingly blue eye, some cavorting, some hiding, most seeking each other's company. As she walks us back into the first building, the indoor kennels, Jerilee notes that the motto of TRHS is "Where love finds a home." There's a bag of little packets of earplugs at the door. I wave it away, but soon understand that I'd be using them if this were my daily occupation, because the dogs are barking, and their improvisational call and response is impressive in both vigor and volume.

Leaving the indoor kennel area, Betty tugs on my sleeve to direct me to a room where the hundred dog bowls are cleaned and sterilized every day. Her fascination is to notice how things work and to think of more cost-effective or efficient ways to get things done. She knows machinery and she loves it. Race cars light her up. She lays eyes on a tractor and asks, "Now is that a double axle or a single axle? Because the angle of that thing. . . ." Betty helped to fund the new reception building. She proudly shows me the dedication plaque erected in memory of Caroline. We stand quietly together for a moment before the little engraved brass plaque in the midst of the chaos of barking dogs.

Then Betty grabs my elbow and steers me in the direction of the project that really seemed to spark her, the purchase and renovation of a motor home to create a mobile spay and neuter clinic. The motor home's insides have been stripped and replaced with a well-equipped sterile surgical area, the walls lined with crates, the shelves stocked with bandages and medicines and extra blankets. The shelter pays a daily fee to a local veterinarian who comes out to spay and neuter, doing between fifteen and thirty surgeries in a day. As they show me the refurbished motor home, I am impressed with the efficiency of this system.

"My ultimate goal is to have a vet on staff, and to get a grant to take this out to the Warm Springs Indian Reservation, where most of our puppies come from." Jerilee emphasizes that they can't just go out and scoop up dogs on the reservation because it is a sovereign nation, but she can offer to help people who want to spay and neuter animals. Again, there is the issue of knowing and respecting jurisdictions that often have differing legal requirements. Betty pipes in, "The last thing, which we don't talk about it yet, is to do education, go to the school and show the little kids this unit and talk to them about neutering."

I can't help but want some of the dogs that I see in kennel after kennel. The puppies are so cute, and the elderly dogs with their frank expressions especially pull on me: what they want is so simple. I can't imagine how I would survive in this world where every time you turn your head, you gaze into another pair of pleading eyes and you feel another tug at your heart. It takes a special breed of human, I think, to do this. And there is something in the work, in the very stark directness of animal needs to be healed and fed and wanted, the something that people refer to when they say that saving other animals has saved them. But saved them from what? I want to say that perhaps many of us need to be saved from the bullshit factor of human living, the pretenses, the cover-ups, the endless analysis and explanation for things that are really simple, like the need for food, for body contact, for healing of life's wounds.

And here is why my grandfather's painting of the little boy and his newly returned dog, and the healing of his grief, so gripped me. My own life hinged on the moment when my father killed a beautiful dog whose muzzle was just turning white, who had been a second mother to infant me and toddler me, who had lived beside me as I entered into kindergarten and then grade school. She was a breeding dog who had let me snuggle with her puppies in the various nests she made for them in the grass, behind bushes, under the house. Sherry (aka Lady Bright III) was a champion beagle who had outlived her breeding capacity, but not her calling. She kept taking the puppies of our younger breeding dog, Winnie, and trying to nurse them. For this supposed sin, my father killed the kindest being I had ever known.

That was a repressed memory that came bursting open within me in detailed clarity after decades of repression, the kind of event whose reality I, a clinical psychologist, did not much believe in until I experienced it. I had long been able to pinpoint the exact moment in my childhood when I was suddenly overtaken with depression after being a happy and successful grade-school kid. And I had written about Sherry's goodness in my book *Spirit Unleashed*. But I had never been able to connect the dots. Until one day when my mom and I sat at the kitchen table talking about the book, and about Sherry. She asked, "You know what happened to her, don't you?"

"No," I said. "I don't remember."

"Ben killed her."

And when my mother reminded me, I burst into wailing grief as the vivid memory of my eight-year-old self slumping over in the passenger seat of the VW bug crashed over me. I wailed all through that night. I could not stop sobbing, so many decades after the deed that had set the direction for my life. Something in me had died in that childhood moment, and that something was painfully resurrected these several decades later. I had tried to save Sherry in my childish way. I insisted that I go to the kill shelter with her. My mother brought her papers, attesting to her champion standing in hopes they would find her a home instead of killing her. But they did not want her papers. I understood the meaning of that, and I knew that I was utterly helpless to save or even comfort my beloved other mother. As I sobbed through that night of recovered memory, I begged Sherry's forgiveness, I pledged my love. In half-waking visions and dreams, Sherry reassured me, but still I pledge my love, still I grieve. As a child, the soft animal of my body loved what it loved without explanations and justifications, without excuses, and that love was formative.[18]

I had fallen into a deep depression on the day Sherry died, and on that day my childhood was derailed. I had been an honors student and the first one chosen for athletic teams. I was the kid who belted out popular rock songs and led the practice of dance steps. But that day I became the kid who sat on the bench staring into space with her baseball glove and schoolbooks next to her. Many years later, in the midst of a normal adult episode of depression, I remembered that deep childhood depression and I said to myself, "Yes, but I came out of it. How did I come out of it?"

My teacher had sent word home that something was wrong with Anne, that she wasn't doing her homework or participating in class. My father's response was to become suddenly unhinged and to beat me viciously. Stunned by the unexpected violence, I was thrown to the ground, hit my head on the corner of a cabinet, and then was pulled to my feet by my long hair. I was kicked across the room. I lost contact with the details, it went on till it stopped. I'd been beaten for being heartbroken that he'd killed someone dear to me. And, on that day, in response to that beating, I discovered my capacity for contained and calculated rage. Rage, the agent of immediate control, rage the correlate of substance P, rage bottled and portioned out, rage effective. It was the decision to stoke and direct my rage that brought me out of depression.

I decided on that day that I would live to fight the bullies, to interrupt their callous, arrogant, and cruel wrongs, and to offer tenderness to the injured beings. I would learn to turn the bullies on themselves. I would learn complete self-mastery, and I would use my newly bottled rage to get immediate control, so that events like Sherry's death would not happen with me in the room. And I have to say that I have interrupted many would-be violent scenes over the course of my life, from the bullying of weaker kids on the playground to the protection of the runts of various litters: cringing wives, innocent animals, terrified runaway teenagers. These were not bullied, or raped, or drowned in the toilet on my watch. No such thing ever did happen again with me in the room.

No wonder, then, I had decided to try to help little Bronwyn. I was like her, but more fortunate in that I had been socialized by two good mothers, one dog, one human—and I was socialized in the company of my many human siblings and all those puppies. I had lots of oxytocin to mediate my large portion of substance P. And every dog I have rescued and loved has been an apology to Sherry for my helplessness, and a thanks to her for making a human animal out of me, one with the basic dignity to honor my connection to other animals, to fight for that very real love. "Anger and tenderness, my two selves," as the poet Adrienne Rich declared, substance P and oxytocin.[19] I am fighting still to wake humans to that connection to our larger family.

Ah, here we see the spring, the source of the emotion that is evident all through this book. Isn't it all about the author's feelings, then? And doesn't that undermine scientific credibility? Yes, the author is motivated by intense—in fact, intentionally intensified—emotions. No, it doesn't undermine the scientific credibility of the work. Affective neuroscience affirms the simple and central fact that all animal behavior is affectively motivated, that these fundamental feeling states express the values that drive and energize us. And this means that the world of animal life, yours and mine, Sherry's and Jason's and Jabulani's, runs on those basic biological values that we have been visiting: curiosity, fear, rage, care, lust, grief, play. The world is not a cold and calculated place in the manner of that old science fiction that humans are the singular rational beings, able, with great discipline, to be unmoved by their own emotions. That is a human fantasy that especially appealed to the early modern philosophers, but it was a fantasy then and it remains a fantasy now. The idea of cold and disciplined rationality is preposterous because we wake with, to, and in emotion, or we never wake at all.

That Sherry and I are bound by love, that I am motivated by that love, and by the grief of her death, and by rage at my late father's socially sanctioned cruelty to her and to me, that I am motivated by these feelings is the normal stuff of animal life on Earth. Of course, animal scientists love animals, and astronomers love stars, and geologists love rocks. I love dogs because dogs first loved me, and the love of dogs has given me a strong core and boundless love for my connection to all of my animal family on this planet.

From my perspective as a philosopher and a psychologist, this love for life beyond one's self is the primary spiritual disposition: it is the motivational place from which we can best live. Because, contrary to the old culture of science that imagined itself rationally above all those messy feelings, it is feelings that make the world go round. Anyone who still wants to argue with that will have to argue with billions of years of evolution that has imbued animals with biological values, expressed as feeling states.

When it comes to dogs and humans, there is a deep affinity of animal species who are similar in important ways, who live socially complex lives, who share bonds of affection, who are best when they love each other. Thanks to the love of a dog for me early in my life, and the many other dogs with whom I've shared my love in this life, I have had a spring of lifelong resilience in knowing deeply and beneath words that I belong to an extensive animal family.

CHAPTER 9

Animal Ambassadors

I had turned off the coast-to-coast walking path that traverses northern England, sixty-four miles in, and I was obsessively anticipating the cake and
cup of tea promised at Lockholme B&B. I was also anticipating a quiet rest
day in Kirkby Stephen after the demanding hikes across the mountainous
Lake District. Chrissie, the owner, greeted me at the door and invited me to
take off my muddy boots and to carry them, along with my pack, to a back
porch. From there, she escorted me to her glorious English garden, flowers spilling over stone walls, where I sat down to enjoy my cake. Glancing
around at the old stone houses nearby, I caught sight of a splash of color on
the dark slate roof of the house next door. I turned my head and saw that a
large red, blue, and yellow tropical bird was using its claws and hooked beak
to work loose a piece of tile. The bird was soon joined by another. I put down
my fork and took my field glasses from my cargo pocket.

Chrissie returned, teapot in hand, and I pointed in amazement, asking for a reality check. Was I really watching Amazonian birds casually dis-
mantle a slate roof in northern England? She laughed and told me that she
and Joe, relative newcomers to Kirkby Stephen, had not quite grown accustomed to them, but that the townspeople had taken them on as a kind
of town mascot. She gave as evidence that the football field, named Parrot
Park, had been donated to the town by John Strutt, who had introduced the
birds to the area, and at Parrot Park a giant painting of the macaws now
announced the town's embrace of them. She explained that a flock of macaws lived in open aviaries on a parrot conservation reserve in the Eden Val-

ley, directly adjacent to Kirkby Stephen. Further, she reported, residents of the town were not only brightly represented but also insured against damages caused by the macaws because these birds were endowed by the same John Strutt.[1] The kindhearted man who had trained them to be free-ranging birds in an environment very different from that of their native land had also cared for his human neighbors, and cared for the relationship between the humans and the birds.

Dropping by the visitor center in Kirkby Stephen, I found a good many scarlet macaw icons for sale: refrigerator magnets, postcards, tea towels, paintings. There were also a couple of small publications describing the history and current status of the Eden Valley conservation project as a kind of home away from home for the large, noisy, and colorful birds. As I was leaving with reading material in hand, the volunteer pointed me to a poster on the door. The annual open-house garden party at the Eden aviaries was scheduled for my one day in town.

It was a sunny Sunday, late morning, when I walked through the open gates of the estate that houses the aviaries. I made a small donation to the charitable trust and asked about its mission, learning that there were actually two trusts with related missions, the conservation project, totaling about thirteen hundred acres of farmland returned to the wild, and the Centre for Parrot Conservation. I asked with some excitement to see the open aviaries, explaining that I write about human-animal relations, and I was given a guided tour.

The aviaries themselves were large wire cages in rows, opening to a path down the middle. On the upper floor, each compartment had two openings through which the birds flew freely, a front door and a back door. I learned that they went out in the mornings to roam the town and countryside, and returned of their own accord in the evenings to eat and to sleep. I noticed their roosting boxes high in the trees above the aviaries.

When I first entered the aviaries, I was introduced to Peanut, an African gray parrot who was swinging gently on a perch very close to the wire-mesh border that separated us. I had long desired to meet an African gray, having known of and been entertained by the antics of Alex, animal cognition scientist Irene Pepperberg's parrot research partner in the United States.[2] In my excitement, I forgot my most basic manners and stuck my finger in to say hello before introducing myself, before receiving permission for greater intimacy. Peanut promptly took a chunk out of my knuckle. Note to self: Do not be so distracted by human companionship that you fail the basic requirements of relating to animals. I have noticed that when I am with other humans, I am less deeply attentive to other kinds of animals, as though I am pulled up to a more superficial but distracting level of existence. Simply put, I don't listen nearly as well, here to my detriment!

My finger bled profusely as I searched my bag for a tissue. I pushed the bruised meaty flap of skin back in place and applied pressure. The aviary keeper continued walking me around the cages as the blood streamed out of my soaked tissue. It was another man, whose suit and tie were paired with a straw hat and elegant walking stick, who finally enquired after my bloody hand. He quickly dismissed me as thick and deserving, but he gladly entertained some of my questions about free-flying parrots. I asked him how they handled complaints from people who objected to the noise of the parrots as they flew around the Eden Valley. "We tell them that the parrots live here and they can go somewhere else!"

When my guide and I arrived back at the group of garden parriers, a woman wearing a cream-colored suit and matching flowered hat asked about the blood now dripping to the ground from the hand I was holding at arm's length from my body. "Good thing it wasn't a great white shark you wanted to meet!" she hooted. I'd alternated between anxious chagrin about my own stupidity in having invited the parrot to bite me, and worry about my inability to clean and close the painful wound, while really wanting to pay attention to the birds and the people who make this remarkable program possible. I was relieved when the comedienne sent someone into the stately home to locate a first aid kit. I chatted with next-generation relatives of John Strutt while cleaning and bandaging my wound. And I learned that Peanut, the African gray, had imprinted on John Strutt himself, even flying into the house to lay an egg under the dressing table in his room. Peanut had often walked the gardens sitting on John's shoulder. I now felt honored to have been chomped by Peanut, and I consider that still-visible scar to represent an honorable lesson in basic manners.

As I listened to stories from people who had known him, John Strutt seemed to me a person who spent his life recovering from a vision of what he was supposed to be by becoming the person he was made to be, step by step, through his own combination of learning and courage. He especially seemed to have discovered a learning of the natural world that affirmed his tenderhearted nature, and at each stage he found the courage to embrace what he had most recently learned.

Strutt was born into a wealthy family, raised to live the life of a country gentleman. His mother and father had engaged avidly in fishing and stalking, the British term for deer-hunting. Indeed, his mother had famously bagged one last stag on her eighty-ninth birthday.[3] His biographer wrote, "John, at the customary age of eight, was sent away to prep . . . and from there to Harrow. There his lifelong distaste for the establishment took shape, and he did not blossom either in sport or scholarship." The one thing John seems to have taken from that wretched high school experience was the kindness of a part-time music teacher, Richard Hodgson. After Harrow, John went on to a tour in the army and then to college, after which he managed a large farm, all the while paying attention to and learning about the wildlife he loved, especially birds. "His sensitive eye for nature grew stronger as the years passed."[4]

John always remembered one particular day, Saturday, May 19, 1956, because on that day he found at his attic window the cock from a pair of budgerigars given to him by his mother, and set free by him. He put the bird who had returned to his window in a small cage on the roof of a shed and waited for the return of the hen, who after two days did indeed return, hungry. She needed food and she was drawn by the sound of her mate's call. John continued to allow the birds their liberty while training them to return to him. He eschewed hunting and fishing, and he learned how to build aviaries and train birds, eventually designing the open aviaries that I saw full of parrots and macaws.

John observed that some birds were bad fliers, unable to navigate, therefore unable to return to where they had flown from; they were birds prone to losing their way. He believed this was often if not always a consequence of their having been confined to cages at critical moments in their development. "They *must* be allowed to fly" when they fledge, he said, or they will become brain damaged. Of his lifelong love for birds and his invention of training methods and aviary designs to support free-flying birds, he said, "I never had a role model. I learnt as I went on. In the beginning I made plenty of mistakes. I was on a learning curve."[5] This sentiment again. It's one I heard almost everywhere I went, talking to people who were trying to save animals, the mantra of figuring it out as you go because there is no beaten path to walk.

Not only did John adapt over time, but the world around him changed. In 1981, the British Parliament passed the Wildlife and Countryside Act, making releases of nonnative birds to the wild illegal. John's birds were from warm climates and thus not able to adapt to the Cumbrian wild landscape. He could cage them, but that would have been an affront to his deep belief that birds need to fly free in order to be birds. Fortunately for John and his birds, the legal decision was that free flight was not the same as release to the wild because the birds had to return to John for their food.

When he first moved to Eden Valley, John instituted the best-practice farming that he had learned in college and had practiced in his previous employment as a farm manager. He plowed the fields of wildflowers and reseeded them. But as his sense of the intricate interconnections of wild plants and animals developed, he had a change of heart about what constituted best practices. He began to restore and create new wildlife habitats, including wetlands surrounded by woods that he also planted, all in imitation of what nature herself had made in this part of northern England, and he put the restored lands in trust as a habitat for wildlife.

John Strutt cared for his worlds, the humans around him, the animals, and the land with its many intricately interconnected and self-sustaining entities. He was a man who seemed to move against the grain of his social world while also honoring its requirements, exercising his capacities to change himself and the world. Many years after his terrible, horrible, no good, very bad days at Harrow, John by chance discovered that his old music teacher was living a lonely life in a care home in northern England.[6] John carried him back to Eden Place, where he cared for him for seventeen years until Richard's death.

I had arrived six years too late to meet John, yet I continue even now to be edified by his story. As I left the open-house event at Eden Place, I stopped under the great trees filled with a cacophony of parrots. Leaning against the iron fence, I texted home once again, "Charlie bit me," with a photo of my bloody knuckle followed by a photo of Peanut. Peanut had imprinted on John, and I had now been imprinted in a different manner by Peanut. I had come once again to understand that some animals live their lives as ambassadors, as creatures who bridge the gap between lands and worldviews, between humans and nature. Peanut had reminded me that respect for his kind was critical to our relationship. Not everything can be smoothed over by an ambassador.

Before leaving England, I had one more visit to make. I made my way to Monkey World Ape Rescue Centre, 350 miles south of Kirkby Stephen. Home to more than 250 primates of twenty different species, Monkey World is housed on sixty-five acres in the Dorset countryside. It was a rainy December day and I had taken a dawn train down from London. After paying my entrance fee, I headed straight to the cafeteria for a hot cup of something. While enjoying my steaming tea, I read the orientation brochure. I learned that the place was established in 1987 "to provide a permanent and stable home for abused Spanish beach chimpanzees."[7] Of interest to me was the fine print in a bottom corner of the brochure: "Please note: this is not a visitor experience." It reads seemingly without reference to anything in particular. So I see it now as I felt it then: the whole show is not intended as a visitor experience. It's not for you, the human. It is for these others, the other primates. This is their place, and you, human visitor, are in their home.

It was definitely the off season, this rainy day, and there were many repairs and building projects underway. The management, having survived the busy season, was not to be found. Perhaps they were attending to other aspects of their mission, like working with foreign governments to stop the illegal smuggling, abuse, and neglect of primates from the wild, and to allow endangered species to breed in selective cooperation with international programs. It was all a bit disconcerting at first, a bit unwelcoming, a bit unnerving, and all of that served to underscore the point that this place was not about me, the early and almost lone visitor. Here no one cared about my money, no one groveled to assure my customer satisfaction, no one offered me a guided tour. I was welcome enough, provided I behaved as required and stayed out of the way. I came to experience that as a deep kind of graciousness. I walked the paths, quietly pausing from time to time to try to catch a glimpse of the various primate families. I saw mating monkeys, monkeys eating, and monkeys playing chase down a tunnel between two buildings, equipped also with swinging ropes and platforms. This was not like any zoo I have seen; it was clearly made from the primate's point of view, equipped for their bodies and their interests.

I was startled, scared for a flash, when a chimpanzee leaned over and smacked the glass window through which I peered while standing out in the rain. He bared his teeth in a grimace before retreating to his bench on the other side of the glass and resuming his philosophical thoughts, chin cupped in hand, brow furrowed in contemplation, leaving me to wonder forever what he was thinking about. In another area of the same building, I watched chimpanzees arranging their sleeping hammocks, some inviting friends for a rainy-day cuddle or gentle swing. Staff members mentioned to me that, had I come on a sunny day, I would more likely find them in their adjacent park with its giant swings and slide. The rain underscored that I was the one on the outside and that it's not only humans who enjoy rainy-day warm snuggles.

My few hours at Monkey World made a unique and lasting impression on me. These were not wild animals, nor were they domesticated animals: these were rescued animals, many of whom had been kidnapped, "neglected, kept in unnatural conditions or experienced unbelievable cruelty."[8] Now they were cared for and supported in a world designed for their comfort, and the difference in the animals was palpable. They lived in family groups. They did not exhibit the mental illness that is unfortunately common in zoo animals, no pacing in circles, no lethargic hiding behind props made to look for the human gaze as though this were a natural setting for them. I was incidental here, and the animals were emboldened enough that one might tell me to stop staring and interrupting the privacy he needed for thinking deeply. This, I felt, is a good place for these particular animals, who cannot be released back to the wild because of what they have suffered at human hands.

On my way out of Monkey World, I stopped to buy a gift. So many primate-themed things to buy, so little room in the suitcase. So I "adopted" Lulu, then estimated to be twenty-four years old. Lulu had been born in a traveling circus in Cyprus to a mother who had attacked her and bitten her arm. Lulu's bite wound had become infected, and a Cypriot family brought her to a doctor who saved her life by amputating her arm. Maybe Lulu's mom would have bitten her arm and killed her in the wild. Maybe Lulu's mom would have been calmer and never bitten her in the wild. We don't know. But, by adopting Lulu, I was also helping some monkey or ape who wasn't here yet. And I wanted to adopt Lulu because it is feeling for another that motivates generosity, and I'd seen Lulu in her family group and not even noticed that her arm was missing. I liked Lulu. I felt sorry for her past and her lost arm, and I was proud of her for living so well in spite of it. If someone on the street had handed me a pamphlet describing the horrors of kidnapping infant primates in the wild and asked for a donation, they'd likely not have got anything from me. But Monkey World, asking for a donation in the name of Lulu? They got it, because that is where we live, in the feeling for one another, and my feeling for Lulu was one of admiration and respect.

When I was a kid, I thrilled at the opportunity to attend a small circus because there were two elephants I could encounter there. I can still recall feeling the immensity of them, their wise old heads, nimble trunks, and the chains around their legs. And I jumped at any opportunity to visit a zoo because I loved animals and wanted to see them, to meet them face to face, to touch them if possible. The same biophilia possessed me as possesses most of us who have curiosity and natural affection for our fellow beings. But I vividly recall crossing the parking lot after a visit to the Minnesota Zoo, now as an adult. I was in Minneapolis for a psychology conference, having recently completed my clinical training. I had been excited to have the opportunity to visit a place renowned as one of the best zoos in the world. This time, however, because of my clinical training, I knew what I was seeing in the lethargy of posture, in the stereotypical movements of pacing, circling, swatting at air: some of these animals were mentally ill. Many were depressed, literally bored out of their minds, and sometimes pathologically lonely. My heart sank as the personal realization of what a zoo is opened up within me.

Zoos historically have been designed primarily for human experience, while the people who staff zoos do so because they love animals. Most zookeepers want with all their hearts for the animals to be happy and healthy, and, indeed, the Minnesota Zoo now offers integrated animal enrichment programs. But the fact remains that the twenty-first-century human species, even the best of us, has had to wake up to the uncomfortable truth that cages and artificial environments—while preferable to endangerment and abuse—are not as good for animals as are their native environments, free and wild. I felt it again when I had a private tour of the primate areas of the San Diego Zoo years later. There I saw a lone Asian monkey, the only one of his kind in the zoo, who sat masturbating in his cage. Young adolescent humans called each other over to watch and to giggle at the poor monkey whose only source of liveliness was his own penis, which he rubbed relentlessly just to feel alive. His predicament and the whole scenario broke my heart; it just wasn't funny.

I am not alone in waking up to what zoos and circuses are—or, I should say, what they historically have been. Animals whom we thrilled to use as entertainment, whom we edified ourselves by observing, whom we ennobled ourselves by studying, we now call ambassadors because we understand that they bridge worlds for us. We no longer think of them as things, but roughly as persons, agents centered in their own lives, beings who have interests and desires, some of which may be incomprehensible to us, most of which likely go unfulfilled. We now recognize that these exotic others experience life as a qualitative event, sharing with us the fundamental desires for self-enrichment, play, social bonds, family life, and sharing with us the experiences of frustration, anger, grief, depression, even the madness of post-traumatic stress, when not allowed to do what they can do, when robbed of their family connections and the capacity to become the skilled animals that their natures define. Paradoxically, because we have come to understand their personhood by bringing them into our world, countless animals are now designated as ambassadors to humans, a step up from the mere entertainment missions of zoos, circuses, and animal-study projects of the past, a recognition of the "someones" they are (I refer here to the ways that wild animals were loved by us humans in the past, not to the many ways in which they were treated as objects, exploited, even tortured).

This change of nomenclature reminds me of one of the vignettes from *The Little Prince*, that of the king who inhabits a small planet, alone.[9] When the little prince, who has dropped in from the sky, refuses to stay to perform the role of subject for the would-be king and instead takes his leave, the king of nothing shouts after him, "I make you my ambassador!" The analogy stops where the little prince is actually able to leave the self-acclaimed king. The other animals, the once wild animals who live in zoos, in research-facility cages, and in circus trucks and wagons, they cannot take their leave. It makes little difference to them whether we call them captives or ambassadors, but it makes a difference to *us* that may play out as truly beneficial for them, because we want to treat ambassadors with due respect.

It's now widely understood that zoos and circuses and animal-behavior research projects are in ethical trouble for their sometimes callous exploitation of animals. Again, I refer specifically here to research projects whose goal is to understand other animals, not to use other animals for research intended to benefit humans at their expense (to my mind, there is no justification for harming other animals to ensure that the fragrance in my cosmetics won't harm me). The meme has changed. These human cultural events have allowed us to get to know the animals. But ironically, the very thing we have valued—bringing exotic wild animals into our midst in the form of urban zoos, circus entertainment, behavioral research studies—we now see as wrong for the animals we have known in such exploitive ways. In an article that both critiques the definition of the traditional zoo and suggests a new role, Lisa Kemmerer writes, "Animals in zoos are used for science, captive breeding, perpetuation of species, education, entertainment, and profit. Ultimately, zoos confine nonhuman animals *for* human beings. . . . All zoos exploit nonhuman individuals for human ends."[10] This is the simple fact, that zoos have existed only by and for human purposes, without real regard for the animals who live there.

Zoos are racing to catch up with this ethical realization. Every zookeeper in a credible zoo is designing enrichment activities tailored to the evolutionary behavioral specializations of each species of animal, and all credible zoos try to make environments that are syntonic for the species they house. Many have built or are building wild animal parks, sometimes putting animals on rotation between zoo displays and time off in the more natural setting of large parks, as is the case with primates at the San Diego Zoo. A simple and sad fact is that there are many more noncredible privately owned roadside zoos than there are large accredited zoos, and there are many smaller zoos that are accredited but lack the resources of large urban zoos. These smaller zoos express various levels of interest and capacity to meet the needs of animals confined within them. While well-meaning zookeepers of all kinds try to provide the right physical circumstances of environmental conditions and diet, it is a much bigger job to provide for the social relationships and enriching achievement activities that other species need.

It was a staff member at the National Zoo who noticed that Shanthi, an Asian elephant, was fascinated by making rhythmic sounds. Debbie Flinkman, elephant keeper at the zoo, explains what the elephant keepers observed. "When Shanthi interacts with her environment . . . she likes to tap on things with the tip of her trunk. She likes to flap her ears against things that make noise."[11] So they affixed a harmonica to the fence posts and installed horns and other musical blow toys "for her own pleasure and joy." Now Shanthi can be heard playing her own compositions, each of which has "a beginning, a middle, and an end . . . and it has a big crescendo at the end." In the video, Debbie suggests that the animals here have it better off in some sense than do those who live in the wild, an opinion often expressed by those who care for captive animals. "We offer the elephants opportunities to do the same things that they would do in the wild here at the zoo or park, only they get to do them here for pleasure almost and in the wild they have to do them to survive." I would be there with Shanthi in a hot minute if I could be, but I am watching a video of a captive animal, taking what pleasure she can in her too-small confinement. Would Shanthi rather be taking her chances in the wilds of Asia or playing the harmonica at the National Zoo, in a small place where she gets free food, medical care, and an appreciative audience for her compositions? We don't know. We don't even know how to ask her the question.

Elizabeth Marshall Thomas, the cat anthropologist, makes the case that, while most people easily understand the wrongs involved in training wild animals to perform circus tricks for humans, circuses are likely preferable to zoos for big cats for the simple and powerful reason that in the circus the animals have something to do.[12] She notes that imprisonment is the only condition humans experience that is roughly parallel to a wild animal's confinement in a zoo. Describing the particular situation of big cats in zoos, she asks us to imagine living with a sibling in a department store display window that is made to look like a place of luxury, but in which everything is a façade, the doors open onto nothing, the radio makes no sound or sense.[13] Imagine, she suggests, that you and your sibling have been confined together so long with nothing new happening that you have nothing to say to each other anymore. Rather, your reality is a relentless depth of boredom in the context of being watched by other beings who come and go freely, gasping at your beauty and your lush environment. Recall that cats like to be concealed in hiding places. Big cats in zoos actually try to use their environmental props of rocks and trees and waterfalls as hiding places, where they seek privacy until the part of their day that they like best, when they go inside to sleep confined in a cage, out of view. "You and your sibling lie down behind the sofa, where you escape into dreams. You don't wake up if you can help it, not even when people in the crowd notice your feet poking out beyond the sofa and bang on the glass to rouse you. You dream of the night, which you spend with three or four other prisoners shackled to the chairs in the employees' lounge. At least you and your fellows can talk all night without wild-looking faces staring at you."[14]

The most basic affect of our shared mammalian instinctual brains is what Jaak Panksepp and Lucy Biven call the seeking system. To call it curiosity is not quite accurate because it is the most fundamental drive to wake engaged with the world as one's self. This seeking affect is shaped by the particular body morphology and sensory apparatus of a given being, and also shaped developmentally by lived experience. A baby primate and a baby human likely have a very similar experience of seeking; they wake to their mothers, to their families, to their environments, even to their own bodies. They then take different paths, as the infant primate learns to swing on vines, make leaf nests in trees, gather fruits, and so on, while the little human learns to walk and put on clothes and go to school. These activities are interesting to the little ones. But if they are still doing the same thing after decades, seeking is no longer engaged and they are bored, perhaps even depressed. In fact, the midlife crisis phenomenon is experienced by chimpanzees as well as humans, when the seeking system adapts to a slower pace of novel stimuli and efforts toward behavioral mastery.[15] What is it like to be a primate who grows up in captivity, even in an enriched environment, and how does that compare with being a primate in the wild? Is one midlife crisis as good as the next? Or do depressed animals never get to the midlife developmental event because they were understimulated and underachieved to begin with? We don't know.

Just before writing these words, I went to revisit CALM, the California Living Museum Zoo, the accredited zoo in Bakersfield, California, whose stated mission is "to display and interpret native California animals and plants for education, conservation, and research." The zoo is managed by the Kern County superintendent of schools and it serves as an important part of educational programs, in addition to providing wild animal rescue, rehabilitation, and sanctuary. When I trained there as a docent and wildlife rehabilitator, the goal was to receive only native animals and to have only animals who could not be released back to the wild as inhabitants of the fourteen-acre park that now contains more than four hundred species. In preparation to work in the animal rehabilitation part of the program, I was trained in animal enrichment activities. I like to think that, had I stayed on, I might have discovered uses for a harmonica at the little zoo.

CALM is the place to which I had brought my raven friend Jason. This is the place to which local people know they can bring the foxes and skunks, the raccoons, possums, birds, snakes, and lizards they find injured on road-sides and in fields. This is the place where animal-control officers can bring the errant bobcat, mountain lion, or bear. Here the wounds and diseases of animals are treated, and when possible they are released back to the wild. But when I returned recently to look again, I felt the close press of cages and the very controlled environment that is necessary to keep the animals safe. After all, fourteen acres is not enough for a mountain lion to live as she lives naturally, much less enough to provide natural conditions for more than four hundred other species. Still, on that hot and quiet summer afternoon, I was also impressed by the cleanliness, order, and care that seemed evident.

But, as I write, there are new troubles at CALM. In mid-2019, People for the Ethical Treatment of Animals (PETA) called for an investigation because two fisher cats were believed to have died of flea infestations, and there were accusations that a skunk had starved to death. "The first fisher cat to die didn't get a necropsy from a veterinarian, the U.S. Department of Agriculture inspection report said, but the second cat did—and the animal's necropsy report suggested a severe flea infestation: '... OPENED BAG THOUSANDS OF FLEAS—FROZE BODY OVERNIGHT TO KILL ... External exam: thou-sands of fleas.'"[16] Sharon Adams, curator of animals, responded to the PETA report, saying that CALM takes in more than a thousand animals a year, and that there is no established protocol for treating flea infestations in fisher cats and many other wild animals, so that treatment rests on trial and er-ror.[17] But PETA stated that an animal care volunteer had made notes about the flea infestation and suggested treatment a month before the animals died. Adams noted that the zoo had been short-staffed.

When I take three giant steps backwards to look at what I know of CALM, I cannot escape the complexity of the entire situation. There are many wild animals in the local geographical area and these animals are increasingly pressured by the construction of human habitat. The zoo takes in injured animals and helps the community to form bonds of affection with them. The likely upshot of the investigation will be a better-staffed and more closely monitored zoo that will continue to provide important services for animals and for humans. There is a lot to be sad about in this story, but let the sadness start where it really starts, with the overdevelopment of wild land that pressures the divide between humans and other animals and that frankly is the leading cause of destruction of animals, individually and collectively.

A bright note amid the dismal news of human encroachment on other animals can be found in the case of the mountain lions of Los Angeles. Los Angeles is one of two cities in the world that is home to a natural population of wild big cats (the other is Mumbai, home to a population of about forty leopards).[18] The mountain lions of Los Angeles exist, and that is surprising, but their existence is threatened, and that is not surprising. The primary threat comes from the way that Los Angeles is geographically segmented by freeways. Occasionally a lion is hit crossing the road, but the bigger threat comes from the way they are forced to inbreed, putting even greater pressure on their lack of genetic diversity.[19]

The proposed solution is underway, an $87 million overpass, 80 percent of which is privately funded and 20 percent of which will come from state funds already designated for conservation projects.[20] The overpass will be located at Liberty Canyon, a site chosen to allow connection between formerly isolated groups of lions. The overpass, 165 feet wide, will soar above the freeways and will be made to feel like an extension of the mountains, complete with native plant cover. Designed with berms and hollows, it will have edges that prevent light and sound from the freeway below from reaching the animals crossing the high overpass. The corridor to be built over Highway 101 will be the biggest wildlife corridor in the world, but it will not be the only one. "There are tiny underpasses for salamanders in Massachusetts and on Christmas Island in Australia there is an overpass for crabs."[21] Wildlife overpasses were first introduced in France in the 1950s and they have been popular in Europe for some time. In Italy, highways typically tunnel through mountains, leaving surface ecosystems intact. This is one of the simplest and most powerful things we urban creatures can do to support other animals, especially as they interface with human development, to provide for passages between animal habitat areas.

Vagility is the term that is used to designate the capacity of animals to move according to their own needs, including needs related to food, mating, and seasonal migration. A 2018 international study that followed more than eight hundred individual animals found that the ability of animals to move declines dramatically in areas with human footprints, not only developed areas but disturbed areas. In undisturbed areas, animals move as much as 66 percent more than they do in areas disturbed by human activity, including things like installation of utilities or pipelines.[22] This study focused on the movement of terrestrial animals ranging in size from tiny pocket mice to elephants, and it reinforces the importance of maintaining corridors for the movement of all kinds of animals, from insects to large mammals, for reasons that go well beyond the animals themselves. The animals are needed to move the seeds of plants, for example. It makes clear the importance of maintaining undeveloped and untouched wild lands. The biologist famous for the term *sociobiology*, also the world's leading expert on ants, Edward O. Wilson, has argued that the solutions to current planetary problems cannot be piecemeal. A dedication of half the surface of the Earth to nature, untouched by human intervention, he says, is the necessary and elegant solution to ecological problems.[23] Once again, it is not a matter of saving any particular big five but a matter of the interconnected five million, upon whom all the big animals, ourselves included, depend. The billions of interconnections that bind life together require some space in which to unfold, naturally.

As we pay attention to the other animals in the world around us, our "we don't knows" gradually transform to knowledge and to solutions. We are just beginning to understand issues of vagility. Of the Los Angeles wildlife corridor project, Beth Pratt, the director of the National Wildlife Federation's California branch, said, "When the freeway went in, it cut off an ecosystem. . . . We're just now seeing impacts of that."[24] The Los Angeles project is slated to open in 2023, and it will create conditions that will prevent other mountain lions living the fate of P22, a male lion trapped in Griffith Park who is unlikely to find a mate anytime soon, after successfully crossing both the 405 and 101 freeways. Can we say that P22 is an ambassador animal? I am fairly certain it would mean nothing to him, but it may be important to motivate us, because putting back a piece of the ecosystem could save the Los Angeles mountain lions. And it will certainly benefit other species, maybe insects, about whom we still know too little.

Perhaps the greatest single insight dawning on humanity from so many directions is that we are all beings, so finely and intimately bound together that there is really no substance to the idea of taking care of oneself and one's own as a life-behavioral strategy. Of course, we sometimes must fight for our own lives, we must compete for our time and place in the sun, but ultimately we can't exclude others from our awareness, our efforts, our care without also sabotaging ourselves.

An illustrative example of this is found in Kelli Harding's book, *The Rabbit Effect: Live Longer, Happier, and Healthier with the Groundbreaking Science of Kindness*, in which she tells the surprising story of a research project conducted in 1978 by bioengineer Robert Nerem. Nerem wanted to study the effects of a high-fat diet on the development of vascular disease, using male New Zealand white rabbits who had demonstrated heart disease patterns similar to those of humans. The rabbits were fed a high-fat diet and they developed the expected outcomes of high blood pressure, high cholesterol, and elevated heart rates. However, when it came to measuring actual fatty deposits in the arteries of the rabbits, one group had 60 percent fewer fatty deposits than the other groups. Because he was unable to locate the important cause of this effect, Nerem analyzed and reanalyzed the research protocol looking for clues, and finally what stood out was that the healthier rabbits had all been fed by an unusually kind research assistant, Murina Leveque (I salute you, kindred spirit!), who had petted and cuddled and talked to the rabbits during the feeding process. According to Nerem, "She couldn't help it. It's just how she was."[25] The loving connection, even across species boundaries, buffered the negative health effects of a terrible diet. Who knew? Perhaps especially across species boundaries, the loving connection meant more to the health of the rabbits than did the fat or lack of it in their diets.

Of all the places I visited, and of all the animal-rescue and conservation workers I talked to, the Earthfire Institute and wildlife sanctuary in Idaho, not far from the Grand Tetons, was the place wherein love was so palpably the critical factor in animal wellness. I stepped out of the car on a chilly autumn morning onto mucky grounds around which were scattered various buildings, walkways, pens, and holding areas, some large enough for three bison and a horse. I found my way to an office building and was introduced to Susan Eirich, executive director of Earthfire, with whom I had made arrangements to visit. She greeted me with a hint of reserve, a common response in people who do animal rescue to people writing about animal rescue, because many writers are primarily critics.

It turned out that it's not accurate to say that Earthfire is about animal rescue. Certainly, rescued wild animals are its reason for being, but it is the answer to the question "why?" that is different. Susan told me that the core reason for the existence of Earthfire was to be with the animals, to listen to them, to get to know them, to love them; and not just to rescue or heal them

from their injuries. She reflected that many people involved in animal rescue are oriented to the suffering of the animals, and that their motive is to
want to relieve that suffering. It is a good enough motive, she thinks, but
here at Earthfire the motive is to learn their inherent dignity, species by species, and individual by individual.

Susan suggested that I should first go to visit the animals with Jean Simpson, the animal handler, a happy assignment, and then return to her for a
formal interview. Jean, whose presence is atmospherically feral, explained to
us that the animals had come from a variety of sources, but his speech was
neither focused nor linear. It was more an ambience flowing with ideas and
images, from which I gleaned some information. Some of the animals were
retired film actors whom he himself had trained. Some were rescued from
fur farms, some taken in for treatment of injuries. We met the three bison,
Bluebell, Rosebud, and Nima. I gently played with a silver fox, Loki. I met
the cougar, Windwalker, and the bear brothers, Ramble and Bramble. We
walked past the bears' three-season swimming pool as Jean explained that
they were now getting ready to hibernate, and he showed us the shed/cave
inside the bear enclosure, the place where the bears would sleep through
the winter.

We moved along a complex set of walkways enclosed with fencing between various enclosures. I admit that I had some of the feelings I always
have in zoos and sanctuaries, the feeling of the oppression of wire fences,
cement floors, spaces smaller than some of the animals would inhabit in
nature. But I also felt something else entirely, a certain ease in the animals,
no dullness, no depression, no stereotypical movements, no pacing, rocking,
head-banging. These animals were, for lack of a better description, mentally
healthy.

I returned to the office and Susan and I walked together to a large circular meeting room that is used for retreats. I asked the usual questions and
received some unusual answers. "Because of Jean's particular skill with animals and my particular passion and hopefully skill in communicating to humans, we decided we really needed to join forces and try to communicate
who these beings are to other humans. . . . The animals here . . . are wild, yet
domesticated, and so there is a chance to get to know them quite differently,
and to get to know them over their entire lifetime. The animals here are
semi-wild, we don't try to do anything to change them or domesticate them,
just love them and make them comfortable with people. The next step in the
bridge is Jean, who is half animal and half person, like a bridge or translator,
and then there's me who helps translate Jean and the animals to other hu-

mans." If I had heard this first, about Jean being half animal, before walking around to visit the animals with him, I might have kept one eye on the door in case I felt the need to run. Instead it is an explanation that fits what I have just experienced. And not only that, when I interrupt and offer some excited bit of wisdom, Susan listens with an animal stillness and depth that tells me I am being received, seen, heard. It's a good thing I am not trying to fake anything because that would not work here.

We chatted excitedly, Susan listened deeply, we talked about the change in human sensibility toward other animals, and we agree that it is a big and meaningful shift in human culture. "There is a shift, a huge shift, it's like both things seem to be happening at the same time and there's a race, it feels like a race. What's going to win? Or is it that the old is destructive and the new is waiting to be born and sometimes the old has to get really, really bad in its last throes before the new can be born? I don't know, but it's a nice thought. But the new is being born all around the world." Somewhere in this conversation we laugh simultaneously and say that we really don't need to talk to each other, so simpatico is our perception of the humans, the animals, and the world that we share.

Like other people who work for the well-being of nonhuman animals, Susan tells me she does it twenty-four hours a day: awake she listens, thinks, works, asleep she dreams it. More unusually, she says that she honestly thinks she has not made any major mistakes. "I think in some ways I did not know who they were and therefore you could call some things mistakes, I guess, but it's a matter of spending time with them that makes ever-growing awareness." Susan and Jean are not saving the animals, nor are they being saved by the animals, but they are loving the animals, listening to them, growing in their awareness of what life on Earth is, and trying to help other people do that too.

Susan is a fountain of stories about these animals, beautiful stories of magic and mystery, but I have agreed not to tell her stories. Of these stories she says, "They are all about the animals being exquisite individual beings." And because they are stories of relationship, those stories belong to her and to the animals with whom she communes, not to me. They are hers to tell, and she is well worth the listen. "One of the pleasures of my life here is that I get to live with bears and wolves and bison and cougars and coyotes and foxes and badgers and porcupine and lynx and bobcats and dogs and chickens and geese and deer. And so it is widening, it's so widening." Of Earthfire, I had the most stark takeaway, that these animals are loved for themselves, and that it makes a palpable difference in the way that the animals, and the people, are. Con Slobodchikoff, who has patiently and scientifically decoded the language of prairie dogs, wrote of his own experience at Earthfire, "As a scientist I felt supported. Earthfire is very well aware of science and tries to take a humanities as well as scientific outlook on things. There is some antagonism between science and people who rescue animals. . . . Biologists view animals as populations so they would say Earthfire is a waste of time—wolves and cougars and foxes are not endangered. But every life matters and it is not trivial to the individual animal if they live or die; are loved, hated, or shot."[26]

Perhaps these animals are as close to ambassadors as I have seen. Do they want to be? They seem comfortable and content. But would they choose to be here, given a choice? We don't know.

The same question has often been asked about Koko, subject of *The Gorilla Who Talks*, the 2015 PBS documentary that chronicles the relationship between Penny Patterson and Koko.[27] Koko was born on the Fourth of July in 1971 at the San Francisco Zoo, and she died in her sleep in 2018 at the nice old gorilla age of forty-six. In a *National Geographic* reflection titled "Why Koko the Gorilla Mattered," Douglas Main wrote, "Koko, the western lowland gorilla that died in her sleep Tuesday at age 46, was renowned for her emotional depth and ability to communicate in sign language."[28] I am startled by the use of "gorilla that died" instead of "gorilla *who* died" because, if ever there was a gorilla we knew as a who, as a person, it was Koko, the gorilla ambassador to humanity.

It was in 1971 that Patterson, then a graduate student, began teaching sign language to baby Koko, unwittingly beginning a relationship that came to define both of their lives. Koko lived almost her entire life with Penny in California. At six months old she contracted a deadly disease and had to be separated from other gorillas, to be treated and raised by humans, first at the San Francisco Zoo and then at Stanford University. But the time came when the Stanford project was complete and the funding gone for the gorilla language research project, and the zoo reclaimed Koko for breeding purposes. Patterson, meanwhile, had grown to know Koko as a person, even a daughter-type of person, and a person whose relationships should not be casually severed. So Patterson and her partner, Ron Cohn, established the Gorilla Foundation near Stanford, and the San Francisco Zoo then sold Koko to the foundation on the condition that she have opportunity to mate. So Koko grew up in the company of a male gorilla, Michael, to whom she apparently related as a brother and never mated with. She never had the baby she longed for, in spite of Patterson's attempts to find mates for her.[29]

Throughout her life, Koko expressed intense desire to have a baby gorilla of her own. Instead of her own baby, though, Koko famously had two pet kittens whom she treated tenderly as gorilla babies, nursing them and carrying them on her back. The first one, a tailless cat whom Koko selected and named All-Ball, was hit by a car and killed, and Koko grieved terribly the loss of her cat baby. "Frown, cry, frown, sad, trouble," she signed, and she wept. Finally, Koko signed, "Sleep cat."[30] Koko the gorilla had become an interspecies ambassador, mothering kittens and learning and communicating with humans, using human language.

The *National Geographic* article celebrating her contribution to human understanding summarizes that we learned from her that gorillas have a capacity to learn language roughly equivalent to that of a human child, that she learned two thousand words of English and over one thousand signs in American Sign Language. The article also notes that Koko used her capacities in human language to communicate her emotions to humans, and that one of these emotions seemed to be playfulness or humor. Watching film footage of Koko and Penny, it is clear that Penny learned some ape culture and communication along the way, as she slaps her knees and hoots at Koko's little jokes. As far as I know, though, Penny's ape-culture acquisition has not been written up in any ape journals, and we don't know what level apes would assign to *her* intelligence. Meanwhile, Koko lived a life immersed in human culture, living in a trailer, surrounded by toys and favorite foods and volunteers. She had birthday parties and received gifts. She had visits from many celebrities, and one especially charming interview with Robin Williams, whom she recalled from a video she had watched, and whom she invited to a tickle-fest, during which she picked his pockets and checked the contents of his wallet, and after which they shared a long hug.

Was the life she had the best of all possible lives for Koko? No, it wasn't, in spite of her good health, close relationships, meaningful work, and long life, because Koko never got to realize her gorilla potential. She never lived in the wild and she didn't have a gorilla family of her own, so she seems to us not quite a complete gorilla, a sentiment she herself seemed to feel deeply in her continual longing for a baby of her own. And yet she lived a good life, growing to three hundred pounds and five feet tall, having loving relationships, and the job description of ambassador. These things, love and health and work, are the markers of a satisfying life for most animals.

Watching the documentary, you can feel the complexity of it, and the regret in Patterson's voice as the film closes. Penny thinks that in the long run, Koko's capacity to learn sign language and to communicate with humans is actually the less important aspect of the whole project. "The fact that Koko can love, that we can love each other, though we are different species, it gets people thinking deeply about life. That's what we need to do." Patterson's voice becomes almost inaudible on those last words, drifting to silence. Koko, the gorilla who learned human language and culture, lived a full life, and in addition had caused at least some humans to reconsider their idea that only humans are persons.

I wonder if Koko had a midlife crisis, if somewhere in the middle of her life she became less interested in learning new signs for communicating with humans, if her favorite foods and birthday gifts became same-old, if she longed for some other kind of fulfillment. But we don't know. We don't know what Koko would have been in the wild. And we don't know what to do with so many captive animals. We do know that Koko was indeed an ape ambassador to humans, but we don't know if she wanted to be an ambassador.

There is so much we don't know that we might begin humbly by paying attention, questioning our assumptions, not allowing ourselves to feel justified in heated accusations since every one of us is implicated in the mistreatment of animals. And every one of us can be equally spellbound by the beauty of the other animals, can be held in the dignity of their gazes, edified by their abilities. The interface of humans with other animals, whose eyes we can look into, whose voices we can hear even when we don't quite understand what they say, this is a place of natural motivation to care for the world beyond human culture and human technology. Here we are invited back to our more naked selves, to our more basic needs, but also to our own inherent wholeness. Here, caring for them, listening to them, seeing them not for what we want from them but for themselves, we can perhaps remember who we are, too, face to face.

CHAPTER 10

The Greatest Story on Earth

"Well, I'm a rescuer," responds the painter who is working at my home when I tell her that I am writing about human-animal relations. Before I can even ask, she continues, "Yes, they save us. They fill a hole in the soul." If she were a singular oddball, her comments would be unremarkable, but she is one of a growing cadre of people all over the world expressing a radical shift in perspective on human-animal relations. As we have seen, there now exist countless animal-rescue, conservation, and sanctuary programs in continent after continent, country after country, programs that are expressions of a changing global ethos of what it means to be human in relation to the other animals.

The effects of deep-level change—in fact, the rejection of an old meme and the adoption of a new meme in its place—can be seen in the arts and sciences, in philosophies and religions, in business and in law, and in common folk practice. In short, this rescuing, this saving of other animals, represents a radical and broad change in human culture.

It has been less than fifty years since the common assumption in many parts of the world was anchored to the idea that the other animals are not persons with an inner sense of themselves, a memory of their life events, or a sense of being someone, even as they change from child to adult to old age, even as they move from one environment to a different one. Most people who have ever lived have probably assumed that other animals are legitimately the property of humans, to be used to make human life easier by doing some of the work for us, more interesting by entertaining us, and even possible by supplying their flesh as food. The broadly accepted narrative of human progress during the twentieth century promised that technology would increase meat production and assured us that our cosmetics and medicines were safe, having been tested on animals in laboratory settings. It certainly did not include a strong storyline about animals as persons in their own contexts. Until very recently, dogs and cats were routinely and without question treated as things to be owned. Rodents were pests, disease vectors to be poisoned. Goldfish were carnival prizes. Wild animals were captured and brought to zoos and aquariums as exotic exhibits.

But our human sense of who the other animals are and of what we are in relation to them has been changing rapidly and radically over the past few decades. We see it every day in the closing of businesses that exploit animals and the emergence of new ones that serve them, and in the reinterpretation of the most basic principles of law, including what it means to be a person under the law. The new sense of kinship is perhaps nowhere more visible than in those places where humans work directly for the well-being of other kinds of animals. Interspecies relationships are established as a matter of course in the burgeoning animal-rescue, sanctuary, and conservation sector; in the Baja lagoons where the Pacific gray whales birth their young; at the Hoedspruit Endangered Species Centre where the cheetahs are bred and released to the wild; in the English country gardens of Kirkby Stephen where the macaws fly free but come home to roost at night; at Farm Sanctuary where the fortunate resident farm animals live out their days in peace. In these places, people who do the work say three things repeatedly: that the animals, whom they view as persons of various kinds, saved *them*; that the work is a learning process of trial and error, of risk and reward, of paying attention, listening to the animals, talking to each other, and often making up what to do in particular situations as they go; and that the work is absolutely necessary.

In 2017, "The Greatest Show on Earth"—the circus of Ringling Bros. and Barnum & Bailey—closed its doors. Ticket sales had been plummeting following the retirement of the performing elephants, according to the announcement on the website of Feld Entertainment, the holding company. The feeling of magic once elicited by the power of human trainers to make large and fierce animals do their bidding is now overshadowed by the knowledge that these animals were not born to perform tricks, much less to camp in cages or travel on trucks. Now a sufficient number of humans want something better for these animals: an environment that supports their being what their various natures dictate. The circus has been eclipsed by a new overarching ethos, a movement beyond fascination and toward empathy for and identification with nonhuman animals. People want the elephants to have fulfilling lives, or at least not to live what they imagine as tortured lives. The fascination is still there, but its boundary is set now by recognition of the animals as beings in their own rights, beings entitled to live as well as they can on this rapidly changing planet that we all share. What we don't really know is what it is like to be an elephant, to feel happy as an elephant, though we can easily imagine some components of that, like health and good food and safe, comfortable rest and something interesting to do. But we don't fully know what is interesting for an elephant to do. This is still a new kind of question for us, and one that shows the breadth and depth of our changing perceptions.

I like to call the broad and deep change in human culture the greatest *story* on Earth, a story of humans coming home to their extended animal family. For many of us, the magic is still there but the tent has gotten bigger. The tent now is Earth and we understand that the quality of the relationships has to change: we humans can't continue as masters but must imagine into existence a reciprocal way of being, something perhaps more ecologically correct. It is obvious that the sense of kinship with other animals is not entirely new; it has been reported as a view held in indigenous cultures that date back millennia. And there have always been individuals who diverged from their culture in seeing other animals as worthy beings, as well as groups of people—like Jains in India—who reject all violence against animals, including the act of treading on an ant, as an abomination. What is new is that the sense of human-animal kinship seems to be emerging both into dominance and into greater synthesis with other stories within a new global culture. Thanks to photographs from the moon and from the international space station, we can all now picture what Earth looks like from beyond its bounds. Thanks to the internet, we all know what it is to see events in other countries up close, to take the perspective of a planet inside a spiral galaxy, even to imagine the sensations of other kinds of creatures. These kinds of experience create in us a sense of being members of the planet Earth, small in a huge universe, yet teeming with life-forms that are all related by virtue of being here with us, on Earth.

I could understandably be accused of naiveté if I did not acknowledge that this experience of global culture of kinship is emerging at a time when animals are being eradicated—individually and as species—at unprecedented rates, by loss of habitat, climate change, poaching and "sport" hunting, and by industrial farming.[1] The rate of destruction of nonhuman animals corresponds to the rate of growth in human populations, as people turn to other animals in the ways they long have done, only with increased numbers and more efficient technologies. I do not claim that there is a single, cohesive new story but rather a sensibility emerging from many stories into a global movement of people who see other animals as being worthy of attention in their own contexts, and who grieve—and perhaps feel responsibility—for the suffering of animals and the rapid extinction of species. This new sensibility challenges us to change our thinking and our behavior. It makes us question our right to demand so much of the other animals: their labor, their most basic forms of natural happiness, their very lives. There are hard questions asked of us. For example, it would be a rare person who could witness or even hear a detailed account of the abattoir, where animals are slaughtered, who would not have to ask himself if perhaps he should eat differently. It is easier not to know than to take on a big lifestyle change. However, the questions don't only challenge and threaten us—or, frankly, we wouldn't be asking them—but they also direct us to new psychological resources in the rediscovery of our own animal natures. It's like a huge fantasy homecoming for us, the prodigal sons and daughters who are reacquainting ourselves with our lost family and learning that our belonging goes deeper than we knew.

Every day, new stories illustrating this trend cross my desk. I still feel joy in the story of women knitting elephant suits in a village near Mathura, India, the location of the Wildlife SOS Elephant Conservation and Care Centre.[2] The article is illustrated with colorful photos of the women with the elephants, all in beautiful garb. To my delight, the elephants' new robes include gigantic red leggings. Kartick Satyanarayan, the center's director, explains that these elephants have been rescued from extreme abuse, so they are especially vulnerable during this unusually cold winter. Most of the elephants love their sweaters, though some are quick to take them off. Apparently not all cold elephants are the same! The comments beneath the news story speak to readers finding it heartwarming. Yes, this is the oxytocin factor again: that our human emotions are warmed by the very same physiology that soothes the elephants. The trust hormone produced by gentle touch and voice, or even by the memory or image of these, is likely one of the experiences we share across species that causes us to recognize them as being like us.

The biological sciences give us an ever-more accurate knowledge of ourselves, including our human psychology and behavior, and they also give us a better understanding of other animals. Animal linguist Con Slobodchikoff reports an exchange that took place in a meeting of scientists who study animal behavior, in which a speaker reported that a mother bird wanted her offspring to survive and was queried by a scientist in the audience, "Do you mean to say that natural selection has shaped the female's actions to make it seem to us like she wants her offspring to survive?"[3] The questioner represents the mechanistic view of evolutionary biology that assigns mind and agency to humans while denying these same traits in other animals. It was the dominant view within the biological sciences during the twentieth century, contrary to the core principle of evolutionary continuity, which would lead us to expect continuity of form, traits, and behaviors in the animal world and which would preclude sharp distinctions of one kind of animal from all the others. In other words, the principle of evolutionary continuity predicts mind and emotion in other animals. I draw attention to the way that much reductionist biology contradicts its own fundamental principles in order to demonstrate the depths to which the old meme has infused narratives. But the scientific method is self-correcting, and science itself now refutes the conception of nonhuman animals propounded by the objecting scientist.

It would be easy to say that this self-correction within the sciences is the real cause of our changing attitudes, but I think that the reality is more complex, reflecting the interaction of multiple events and perspectives within living and shifting narratives. There was for a long time a double standard in science, and many animal scientists have described it well. Slobodchikoff, who broke the code of prairie dog language by imagining that it might exist and then developing a sophisticated research model to test the idea, describes the two underlying beliefs that dominated and controlled animal science for decades.[4] The first was that animal behavior is always and only programmed by the animal's genes, the mistaken belief that they cannot and do not learn but live on automatic pilot from before birth to death. The second was that animals do not have intentions or consciousness, the now discredited belief that nonhuman animals are not aware of themselves. Notice that these two ideas are not scientific facts but *beliefs* that kept scientists from asking whole categories of important questions.

There is now so much evidence that animals do in fact think and feel and interact intentionally with the world, much as we do, that a major gathering of scientists has concluded that "non-human animals have the neuroanatomical, neurochemical, and neurophysiological substrates of conscious states along with the capacity to exhibit intentional behaviors. Consequently, the weight of evidence indicates that humans are not unique in possessing the neurological substrates that generate consciousness. Non-human animals, including all mammals and birds, and many other creatures, including octopuses, also possess these neurological substrates."[5] This statement is careful and conservative but reflects overwhelming evidence that we are more alike across species boundaries than different, as is predicted by the principle of evolutionary continuity.

The principle of evolutionary continuity is simple and powerful. It states that all life has descended from a common ancestor, with modifications in life-forms that enhance the ability to adapt to a particular environmental niche.[6] Of the many traits shared in our extended kinship, I have argued that "on the basis of the principle of evolutionary continuity, we would expect that large categories of experience, such as fundamental motivations, developmental trajectories, communication, and mental life (including fundamental self-awareness) would be shared across species." Indeed, it is now certainly more reasonable to expect that other animals "think, feel, intend, and communicate than to assume that they do not."[7] In fact, it seems antiquated and ignorant to assume that we animals do not share broad categories of psychology and behavior. Animals used as entertainment have for centuries been good business, where business is defined as profitmaking. But the popular sense of what is good business is rapidly shifting.

The Barnum & Bailey elephants now live at Feld Entertainment's Center for Elephant Conservation, two hundred acres in Florida.[8] The National Aquarium in Baltimore has released its captive dolphins into an ocean refuge, the first of its kind in North America. Marine mammal biologist Naomi Rose, representing the Animal Welfare Institute, said of the National Aquarium's move, "It has serious implications for where the entire industry is going.... I think smart facilities that want to stay in business or get out gracefully are going to pay attention to these societal trends."[9] This way of considering the needs of animals, as determined by their species and even individual temperaments, is a business trend that is likely to grow because it comes from a deep change in human culture. Indeed, captive animals of many kinds are being released from zoos and circuses at a rate that is challenging for animal sanctuaries: the Barnum & Bailey tigers now face the problem of where to call home.[10] The Global Federation of Animal Sanctuaries has accredited 132 sanctuaries in the United States, only eleven of which can house big cats (surely presenting a new business opportunity!).

Also challenging are the cases of animals being released from the confines of scientific laboratories, where there is increasing public pressure for rehabilitative placement. Scientific laboratories have long been expected to adhere to ethical standards in conducting research involving animals, but these standards have become increasingly rigorous since PETA famously released footage of the gratuitous abuse of a baboon in a university medical research laboratory.[11] This case, similar to so many involving the maltreatment of animals, hinged on public awareness—not easily attained, but unstoppable once raised.

In this emerging narrative of kinship as it affects the business sector, I also note that the communications industry finds that interspecies relationships make for effective advertising of (unrelated) products. One such advertisement for Android shows friendship pairs, each composed of two different species, romping together to the dubbing of a song about friendship, ending with a chimpanzee/dog hug at the punch line, "Together, Not the Same." In an ad developed by Optus, a communications company that wanted to convince customers it was worth its higher price, members of an orchestra are filmed as they are taken by boat to a barge in the ocean, where they gather to perform music that is humanly composed whale song. The viewer watches as large microphones are dropped into the ocean while the musicians begin to play. We hear distant whales vocalizing in response, and we see the faces of the musicians full of wonder as the whales draw near and leap in apparent joyous response. It is this that is most noteworthy: we humans feel joy when we communicate with members of other species (as advertisers understand). From another angle, the advertising imagery for Animal Crackers was revised in 2019 to remove images of animals in circus cages as the norm. One news report on the change opens with these words: "After more than a century behind bars, the beasts on boxes of animal crackers are roaming free."[12] We can tentatively conclude that the new meme of kinship is a powerful psychological resource for humans and for entrepreneurs ready to work in more respectful ways with animal others in creating new avenues of business.

Negative public response—to mistakes, misunderstandings, instances of ignorance, and intentionally inflicted wrongs—has impacted business practice, just as, positively, do cute baby-animal images and warm interspecies interactions on film. But human behaviors that were until recently seen as normal and even good are now perceived as wrongs inflicted on other animals, a change in public perception so powerful that it has also brought about changes in the law, even in fundamental principles of law. Take, for example, the iconic case of Tilikum, the lead orca represented in a suit against SeaWorld claiming violation of the thirteenth amendment to the Constitution of the United States, which bans slavery (*Tilikum v. SeaWorld,*

2012). Most surprising at the time was the fact that the case was heard at all, though perhaps not when considered in the larger social context, including the 2012 meeting of the American Academy of Sciences at which a session was held on a Declaration of Rights for Cetaceans.[13]

The case was decided in favor of SeaWorld. In the words of Judge Jeffrey Miller, "The only reasonable interpretation of the 13th Amendment's plain language is that it applies to persons and not to non-persons such as orcas." But the judge offered no definition of personhood, and many scholars in the area of defining personhood, myself included, disagree with him.[14] The task of defining personhood is actually a philosophical issue, and a very complex one, given the history of Western philosophy that is the foundation for our system of law. Philosopher Mark Rowlands has taken on the task of harmonizing what we now know about the inner lives of other kinds of animals with the history of our philosophical traditions.[15] Ultimately, this philosophical work is necessary for making the deep changes called for in our human legal systems.

Meanwhile, SeaWorld continues to keep captive cetaceans in viewing tanks, but I am not alone in thinking that it should follow the example of the National Aquarium in creating an ocean refuge (despite public pressure and revenue loss, it continues as of 2019 to declare that ocean pens are not an option under consideration). In late 2019, TripAdvisor decided to cease selling tickets to SeaWorld on the basis of negative public opinion. And I note that when Tilikum died in 2017, his death—and the corporate profit associated with him—was referenced in an Associated Press piece that appeared among other places in the business section of the *Washington Post*, with the reflection of a former employee: "He lived a tortured existence in captivity. I think all the whales do, but if you had to pinpoint one of them, hands down I would say Tilikum."[16]

A year after *Tilikum v. SeaWorld*, the nation of India took the opposite stance to that of Judge Miller, conferring on dolphins the status of nonhuman personhood and prohibiting their capture.[17] In Argentina in 2014, a panel of judges unanimously agreed in favor of a writ of habeas corpus for Sandra, an orangutan, who was found to qualify for the designation of non-human person.[18] Increasingly, cases of wrongful captivity are being heard in international courts and several nations have banned the use of great apes in medical research.[19]

There is of course an inherent contradiction in the foundation of laws related to animal welfare. Peggy Cunniff and Marcia Kramer note that law as applied to animals has historically had as its foundational concept the notion that animals are the rightful property of humans.[20] New scientific data and a new human ethos powerfully challenge the belief that personhood is an exclusive attribute of humans, and the right of humans to own other animals—resting on such exclusive personhood—is thereby challenged. Law, however, has been developed and articulated in the human context to serve human purposes, ironically causing the movement for the recognition of animals' rights to be couched in custodial terms. It is hard to imagine that non-human animals would freely consent to our social contracts, so it seems that it must remain for humans to represent animal interests in human courts. However, political theorists are proposing models by which animals, especially domesticated animals, may participate as stakeholders in interspecies communities.[21] In case after case, the frontier seems to be how we can make real our efforts to include them in "our" world, or, alternatively, how to redefine our world in terms of their worlds.

In spite of the knotty problems with changing legal systems, in 2019 California became the first state in the United States to prohibit the sale of new products made of animal fur.[22] Other legislation signed into law at the same time prohibits the sale of lizard, hippo, caiman, crocodile, and alligator skins, and also the use of most animals in circuses.[23] And just one year earlier, in 2018, California had passed sweeping reforms for agricultural animal welfare.[24] Though some people might point dismissively to the progressive bent of California, it is very much to the point that California is the fifth largest economy in the world, having surpassed the United Kingdom in 2018.[25] There is no doubt that a new sensibility of human-animal relations is changing in the direction of increased animal rights and increased animal protections, even as the philosophical issue of animal personhood is being redefined by philosophers.

An illustration of the challenge involved in understanding other animals from their own perspective can be found in Mark Tansey's 1981 painting *The Innocent Eye Test*.[26] It depicts a scene that at first looks like it was published in a mid-twentieth-century high school science text, portraying a cow standing before a very realistic seventeenth-century painting of life-sized cows. The cow gazes into the painted face of a cow, surrounded by men in suits, one even in a white coat, holding a clipboard. The men gaze at the cow as if from outside the scene, when of course the painter has placed them clearly within our (the viewers') gaze, all together, as if to say that no one is on the outside looking in, but all are participants, and all together are here for our consideration.

Animals have of course long been featured in human art. Many millennia before the internet was populated by cats engaging in antics appealing to human viewers, the caves of Lascaux featured human depictions of other animals. But artists are newly exploring the portrayal of nonhumans as subjects in their own world of interests, rather than as objects in the world of human concerns.

There has been a similar emergence of nonhumans portrayed as protagonists in literature. Allison Baird's adaptation of *Moby Dick* for young adult readers, *White as the Waves*, telling the story of Captain Ahab from the point of view of the whale, a real historical character, is an early example, and Barbara Gowdy's *The White Bone*, in which all of the characters are African elephants, a more recent one.[27] When I read Laline Paull's *The Bees*, wherein all of the characters are bees, how surprised I was to find the story of a bee protagonist to be a page turner![28] And I confess that I read Barbara Kingsolver's *Prodigal Summer* muttering to myself, "Why doesn't she just make the coyote a character?!" The answer to my question, as all who have read the work will know, was that she was doing something much more subtle in walking us through the very changes of perception that are happening between humans and other animals, until the coyote is a fully realized character whom we are fully able to recognize.[29]

Other writers explore the complexity of human relationships with other animals, telling stories in which the other kinds of animals are more than props in human stories. Karen Joy Fowler's *We Are All Completely Beside Ourselves* tells with unforgettable depth the story of a family that adopts a chimp to live with the human children as a scientific experiment. Brian Doyle's *The Plover* and *Martin Marten* each express the lives of animals as protagonists with the interests of their own bodies and ecological niches.[30]

I find myself thinking back to the contemporaneous tales of St. Francis of Assisi, invoked today by religious and nonreligious people alike as patron of animals, and of his storied ability to communicate directly with them (I commend the story of Francis with the wolf at Gubbio). In a recent scholarly review of six major works on the subject of animals in religious perspectives and contexts, Anna Peterson declares that these works are important because "religion strongly shapes popular understandings of animals' characters, values, and proper roles."[31] To return to my earlier example, I note that the Indian villagers' knitting of garments for elephants might be seen as continuous with a long history that includes the worship of the elephant-human god Ganesha and the clothing of elephants in ritual garb at festival times. But there is an important discontinuity in that today's knitters are moved primarily by empathy for the elephants and not directly by human religious tradition with animals as characters.

Not surprisingly, Peterson finds that a central theme that repeats over and over again is "the abiding tension between the animals' symbolic importance and their actual lives."[32] She adds, "It is easy to nod to animal subjectivity but harder to engage animals as actual subjects.... The role that animals play in our lives are not their only roles ... and probably they are not the most important part of their lives.... It is as though what animals really do and what they are really like has no place in the conversation. Perhaps the next stage in animal studies will involve bridging this gap."[33]

This unresolved conundrum of how to change from seeing other animals as objects in our world to subjects in their own worlds is of course what comes immediately after the death knell of the old meme. What, then, is the way to understand *who we are* in relation to who they are? This is the idea to which various narrative strands repeatedly come, that the next step is to develop relationships with other animals in which they—the other animals—help to define the terms of relating, and in which their perspective becomes part of our sense of being "us together" with them. I saw a hint of what this might look like at Monkey World, where someone paid attention to what it's like to be a vervet monkey, a chimpanzee, an orangutan, and designed environments for the healing of individuals of these particular species, while recognizing that release to the wild is not an option for the primates housed there. I saw it again at Earthfire Institute, where the entire intent of the organization is that humans listen to other kinds of animals as unique individuals within distinct species.

And this is what I see happening, that we are finally moving into actual relationship with our nonhuman animal kin, that we are beginning to comprehend that they are others, agents in their own right, with particular desires and needs, thoughts and feelings, intentions, even languages which we are beginning to decode.[34] With this comes the necessarily deep knowledge that we can never really own another. Of course, others can be kept captive—but we have a long history of learning that one person cannot own another, not as a slave owner, not as a husband, not as a parent ... because the other is Other, with their own internal center that is not, in reality, an extension of Me. This is not easy to grasp and even less easy to realize in practice, as any spouse or parent knows. A relationship is something like a dance in which two meet and create patterns that neither would make alone, and the patterns cannot be imposed as matters of form but must emerge in the choreography of selves. As with relationships among humans, it is challenging to create a genuinely shared common experience with an animal of another species, but it can be done by extending attention and knowledge, effort and empathy, allowing our incorrect and partial perceptions to be corrected by the other. And the more different the other is from me, the more I may have to work, to tolerate correction, and to be patient with myself, with the other, with the process.

As a clinical psychologist by training, I have to say that this is familiar territory: it is like the work of psychotherapy, by which healing comes to persons and to relationships. The practice of listening and then trying to learn how accurate we have been in understanding the other is both a science and an art. Doing this with other humans poses particular challenges related to the use of words, how they can have different meanings, how they can deceive even when we mean to reveal, how they can remove us to abstract distances from our bodies.

But real understanding is within the realm of the possible, and so is a similar process of listening across species boundaries. When your dog repeatedly barks and goes to the door, then comes to where you sit in your recliner and paws your leg and looks at the door, you know exactly what you are asked to do. In the scientific literature on communication, the term for this common form, found across species, is the "directed gaze." When my eyes make contact with yours, I turn my head and look pointedly at something, I tip my head to say "go there," or I roll my eyes to say "can you believe that?" Yes, the directed gaze is common across species, so that one is easy. But more difficult communications are possible and in fact are underway, and new creative responses to our new set of interspecies problems are also in progress.

Now there are scientific animal-language decoding projects. Now there are national parks for preservation of other species, and ocean sanctuaries for the well-living of other species. Now there are game reserves, no longer for the purpose of big game [sic] hunting but as dedicated habitat for the thriving of the animals who live there. Now there are neighborhood dog and cat rescues, and now there are sanctuaries for those otherwise unwanted, like deaf dogs and unproductive cows.

There are even entire countries with young constitutions that protect nature and the place of other animals within it. Bhutan has designated the preservation of the natural world one of the four pillars of human well-being, the goal of its constitution, adopted and ratified in 2008. In that same year, Ecuador went even further by granting rights to nature not as a resource for humans but as a living being. Article 71 of the Constitution of Ecuador reads, "Nature or Pachamama, where life is reproduced and exists, has the right to exist, persist, maintain and regenerate its vital cycles, structure, functions and its processes in evolution." One Latin American political commentator wrote, just before its ratification, "Jaguars, spectacled bears, brown-headed spider monkeys, and plate-billed mountain toucans may all just breathe a little easier next week if Ecuadorians approve a new constitution in a referendum on Sunday that would grant these threatened animals' habitats with inalienable rights."[35] So it has become culturally normative for us to think about the other animals breathing easier. This is a new day, or perhaps a returning home after a long journey wherein our attention was focused on ourselves, our culturally defined selves . . . and then, coming home to understand that our cultures rest within and depend upon nature, that we are fundamentally biological and our psychology is dependent upon our biology, set within ecological frames.

This is a remarkable change, perhaps even parallel to the response of the Apollo astronauts to seeing Earth from space, how they suddenly understood it to be alive, to be fragile, to be home. Of this experience astronaut James Irwin said, "As we got further and further away, it [the Earth] diminished in size. Finally, it shrank to the size of a marble, the most beautiful you can imagine. That beautiful, warm, living object looked so fragile, so delicate, that if you touched it with a finger it would fall apart. Seeing this has to change [us]."

Our small blue planet has arrived at last in our consciousness, and we are living in an emerging planetary culture. This has changed humanity, as Fred Hoyle predicted it would when he said in 1948, "Once a photograph of the Earth, taken from the outside, is available, new ideas as powerful as any in history will be let loose." And if that picture of Earth from space impresses on us the idea of our shared life and its fragility, the emerging sense of living with other animals as our kin reflects the feeling of being inside that one living entity, Earth. Our planet is alive, and this emerging story of human and nonhuman animal relations is no mere idea, no abstraction of universal rights, but something we know and share, body to body, expressing our internal sense of knowing and feeling and caring that we are one big interdependent life.

There is no doubt that this is a huge cultural shift, but, as the expression goes, shift happens. Consider the shift that happens in the life of every moth and butterfly you have ever seen. From morning to late afternoon I once watched a caterpillar climb a tall pine tree, climb without ever stopping to rest. That was a September afternoon in the beautiful northern woods of Door County, Wisconsin. This tiny creature spent the entire day climbing, up and up the tall rugged trunk, over hill and dale, then out along a high branch, never resting. He seemed to be feeling some urgency, I thought. But who knows? I've not been a caterpillar. He was certainly determined. When he stopped, I thought he must be exhausted, and then he began to spin. Surely, he felt he was about to fall apart; how could he not? Did he know he was packing an incorruptible set of memories of lessons he had learned? And did he know about the imaginal discs that would transform him to something unrecognizable that would still be him living his one continuous life?

But this is nature's way. There's a time when the chick is too big for the egg, a time when the caterpillar may already be carrying little wing-buds inside its body, when the transformation must happen, or there will be certain death. The ugly caterpillar that spins a cocoon, then goes to sleep in its little hammock and wakes up a beautiful butterfly, is legendary. Its actual process is more devastating, less romantic.

The first thing the little creature who has realized all its wormy potential must do is to create imaginal discs of what it will become. These imaginal discs are a few cells, maybe fifty in each disc. These will eventually organize a large number of cells, maybe fifty thousand each. (Does it know what it's doing, does it hope for wings?) The second thing the little caterpillar must do is digest itself, quite literally give up all form, breaking down into a liquid that will become nutrition for the new form. The caterpillar does not turn into a butterfly, but into caterpillar soup with a few imaginal patterns floating around in the soup. For each body part that will be constructed—wings and legs, eyes and mouth, thorax and genitals—all are held in unimaginably compact form in those imaginal discs that slowly gather up the material of the former caterpillar and organize it into the parts that will organize themselves into a butterfly. The transformations feel like death because they *are* deaths, and every life requires many deaths. We do not know how it happens that death gives birth to new beings or new cultural eras, new versions of ourselves that are still somehow continuous with the one who has died.

A recent series of experiments at Georgetown demonstrated that butterflies remember lessons they learned as caterpillars.[36] How can this be? If you cut open the cocoon at its most liquid moment, you would not see any body parts, no worm, no wings, and you would see no memories of lessons learned. You would see brown soup. But the images and memories are in there, unrecognizable.

I like to take lessons from other creatures, those who do life so seemingly differently from the way I do it. I imagine I can perhaps even radically expand my behavioral repertoire that way, perhaps not all the way to wings but maybe to flight behavior or something equally bodacious in its depth of change, perhaps to a world culture wherein humans and other animals live as kin. So I offer this lesson from a worm for this time when we have to let go of our old way of being, when death is speeding toward us on melting glaciers and extinction of species. The memories of our past, the imaginal discs of our future, they are in there, in the soup, the broken down human cultures that will become nutritional substance for some imaginal potential we don't yet know we have.

It's a story that has been a long time in the making, perhaps ever since we left the garden, to use that familiar metaphor. I think of the constipated goldfish whom someone cared about, reported as a sign of hope and progress in an editorial in a major British newspaper.[37] There's not much to say about it. A man in the English Midlands noticed that his goldfish wasn't well, took the fish to the vet, and paid £300 for a surgery that saved the fish from life-threatening constipation. At the level of world news, it's an absurd story. But reporter Deborah Orr called to mind the days when goldfish routinely lived in small glass bowls on kitchen counters and were flushed down the toilet when dead or merely inconvenient. At a different level, at some place deep in our psyches, the story contains a touch of majesty: it approves that a singular person can be about a singular goldfish for a singular moment, and then it approves for that moment to catch the imagination of others. That's how living things work, how our imaginal discs create new memes and shift narratives. About the goldfish, Orr concluded, "When you think about it, this story has been a long time coming." Yes, like the story of human and nonhuman animal kinship, in which there is something majestic in coming home to find a huge family that you didn't know you had, but one that was there all the time. I think it really might be the greatest story on Earth right now; it certainly is one of the most hopeful.

CHAPTER 11

Might We Yet Swarm?

My first animal rescue was a bee. I was four years old. I can picture the green jar-lid upside down on the kitchen windowsill with some water, some green leaves, a flower, and the bee. I had found the bee on the ground and decided it needed rescuing and so had made this impromptu bee hospital, complete with flowers, wherein I installed the bee for two days and two nights. On the third day the bee was up and walking around, so I took her outside, where she rewarded me with my first bee sting. I knew she would now die. It wasn't a happy ending for either of us. I have been stung by many bees since that first one, each time swelling and itching and burning, and each time knowing that the bee loses her life with the sting.

We humans don't share with insects the same neurobiology of attachment and the fundamental affective states that I have used as a point of comparison with other mammals. There are greater differences between our bodies and brains and theirs than between us and any of the mammals. Our shared common ancestor is estimated to have lived a long, long time ago, about six hundred million years ago. From there, our line of descent went the way of fish who moved onto land and eventually became mammals; theirs went the way of crustacean ocean dwellers who moved ashore and became insects. Yet there is this interesting note that, in spite of the obvious differences, honeybees share more genetic similarity with mammals than do other insects. One of the great surprises to the genome team that sequenced honeybee DNA was the discovery that 762 genes in the honeybee are also found in mammals, though lost in flies.[1]

Whether or not it's in the genes, there's a deep strand of human culture that claims something almost mystical about bees, something that allows us humans to feel connected to the world of insects, and through that portal to participate more deeply in nature's secrets. "I remember when I used to see a bee and go, YIKES, a bee! And now I'm all, oh wow a bee, hi! You OK there? Need anything? Can I get you a drink? A cushion? Wanna borrow the car?" Those words buzz around on social media because they resonate. Why bees but not so much the other bugs, despite the fact that all the five million matter, and not only those who live in the big hive? Because, I think, we think they are like us. And why mystical rather than cuddly? Because they are tiny and they have stingers and they are not like us. Perhaps it is also that the bees have an impact on our senses—our taste for sweets, our eye for the beauty of flowers, our love for the warm glow of a wax candle on a dark night, our ear for the pleasant buzz of summertime. We don't want a world without those things. And perhaps we are also drawn to the profound sociality and industriousness of the honeybees because we share these traits with them. They are communal, orderly, and productive. Oh, and they are feminist and democratic, which is to say that either we have a lot to learn from them, or it is really easy for us to project what we like onto them, or both.

Entomologist Jürgen Tautz suggests that the bee colony is a self-organizing and complex adaptive unit of life, based on a network of communication that creates in effect a single being that is similar to a mammal but is distributed across many individual bodies.[2] The idea of a superorganism is one that keeps coming around in biology, and it can be summarized as a model that emphasizes the social unit of the group, the hive in the case of bees, to understand the functioning of the animals organized within that group. In the case of insects, the idea is that each individual animal serves as a unit of mind, roughly comparable to a neuron inside our human brain, so that it is at the level of the hive and not at the level of the individual that behaviors can best be understood.[3]

If the idea of a single mind distributed across several bodies sounds creatively far-fetched, the stuff of science fiction, consider the fact that pioneering social neuroscientist John Cacioppo understands humans best at the level of social membership.[4] From an evolutionary perspective, he notes, our great adaptation is our cooperative sociality. We are necessarily connected to other humans across our lifespan through a myriad of invisible forces that, like gravity, are ubiquitous and powerful. Cacioppo stresses that social species by definition create emergent structures that extend beyond a single organism. For humans these emergent structures commonly include such entities as couples, families, nations, and cultures. Since these social structures coevolved with our biological organismic structures, they together can be seen as our niche of adaptation, our survival mechanism. In Cacioppo's words, "Our survival depends on our collective abilities, not on our individual might."[5] To put it in simple and true terms, a honeybee cannot survive outside the hive for long. In a parallel form, consider the words of developmental psychologist and pediatrician Donald Winnicott: "There is no such thing as a baby . . . [there is] a baby and someone."[6] Cacioppo is quick to point out that this "and someone" is a complex and lifelong condition for us humans, as it is for honeybees. It may well be that there is a depth of social belonging shared by humans and honeybees that can give us some capacity to imagine what it is like to be a bee. But our capacity to understand bees, and (dare we even ask?) to be understood by them, might well take a different form and sensibility than does our capacity to empathize accurately with other mammals, or even with birds, though we can perhaps take a lesson from these others in how to extend our capacities for understanding to ever stranger creatures.

In the mid-1990s I was part of a cadre of young clinicians and researchers in the nascent field of psychoneuroimmunology, now more commonly known as mind-body medicine. The revolutionary neuroscientific discoveries of that era gave credibility to the now widely accepted understanding that the mind and the body are a single interactive system, that thoughts and feelings and brains and bodies are all interactively engaged. One of the pioneering scientific findings of this field was made by Candace Pert in her discovery of endorphins, endogenous morphines, present not only in the brain as neurotransmitters but also in the body.[7] With this discovery, Pert established that the body—not just the mind or even the brain—is sentient, in direct contradiction to the model established by Descartes in the seventeenth century, the model that provided foundational assumptions for the entire modern scientific project. And, very much to the point of this book, it was that model that established an assumed unbridgeable divide between humans and other animals. It is perhaps both ironic and deeply satisfying that science itself, by its self-correcting method, has overturned the assumption of a great divide between us humans and all the other animals, and also now that of a similarly rooted great divide between our human bodies and our human minds.

I recall attending a conference at which Pert was a speaker, and I remember vividly her response to a question about effects of an intervention on mortality by saying that she could not extrapolate because, by definition, her research subjects, the lab rats, all died. Someone laughed. Someone else took offense. Someone else began to make the case that the rats were sentient beings, and another responded that so were snails, and should we just let the snails run over the garden on the basis of that? There is this question lying behind and within our biophilia: how far does it go, and isn't it ultimately impractical? If self-interest drives evolution, we have to compete with all these life-forms to stay in the game. If we grant them all metaphysical equivalence, how are we then to live?

I suggest that the bees may hold some part of the answer to that question, or that perhaps in some kind of interspecies partnership with them we may find our way forward. With the imminent arrival of that time humanity has so long sought, the moment when we conquer nature, we fear intensely that we are losing everything we love, our planet, our home, our extended families, our human family, our very individual lives, in an irrevocable ecological free fall of our own clever devising. We understand that, if there is a way out of the mess we are in, it won't be found by doing the same things we did to bring us to this moment. We will have to find a new way to go somewhere entirely different from where we have been. Perhaps, like the bees, we will find a way to swarm because it has become too crowded in here, because we need a new home in nature, because we need a new way of moving together.

Consider the swarm. When the number of bees in a hive gets unsustainably large, the queen may fly away with thousands of the bees, leaving a daughter to become a new queen in the old hive while the old queen and her newly defined family find another place to live. After the swarm leaves the hive, they surround and protect the queen in a temporary location while scouts go out in search of a new site in which to make the home of the newly reconstituted hive. "When 10,000 honeybees fly the coop to hunt for a new home, usually a tree cavity, they have a unique method of deciding which site is right: With great efficiency they narrow down the options and minimize bad decisions."[8] Their technique includes coalition-building until a quorum develops. The bees use their famous waggle-dance language for more than one purpose when one bee, typically a scout, wants to communicate information to many bees, as when selecting a site for gathering nectar or for establishing a new hive. The strength of the individual bee's dance communicates that bee's assessment of how good the new site is for its intended purpose. In contemporary human terms, a good scout is an influencer. Other scout bees decide whom to follow on the strength of the dance, and soon a coalition forms that favors one site over another, as more bees follow a given scout and report back in their own waggle-dance assessment.[9]

Anthropologist Michaela Fenske, noting the fear among the middle classes of contemporary Western societies that nature is falling silent, says that "the relationship between humans and honeybees has changed from a rather conflictual relationship toward an increased interspecies mindfulness."[10] Perhaps it will be in this interspecies mindfulness that we learn how to swarm, to find our way to a better place within nature. But first we have to stop killing the bees so that they will be there for us to learn from.

Living on my small organic farm in Puglia, at the top of the heel of Italy's boot, I have remembered again that I once lived in a world where there were insects. Not so long ago ants formed assembly lines toward and away from things sweet and sticky. Summer flies buzzed around inside houses and every house had fly swatters. Tiny gnats swarmed the air outside, woodlice that I knew as sow bugs plowed the gardens; earthworms, slugs, snails were everywhere. All these insects were creatures I gradually saw less and less over decades of daily living. You don't really notice the absence of insects day to day, year to year, when your life is about commuting and office work. Now that I am in a world with insects again, I realize there has been a whole lot of insecticide in use virtually everywhere, in cities to keep buildings bug-free, and on farms to keep insects from destroying crops.

I see afresh that we need to learn how better to live with insects because insects are at the bottom of the food chain on which the upper and smaller echelons of the pyramid depend. I've been thinking a lot about insects of all kinds, not only about the bees who are newly "mine" in their homes that I've recently installed, one blue, one gold. And I've concluded this: we've got to become a lot smarter in how we live with bugs, in how and when and where and *why* we kill them, a lesson we should have learned from the evolutionary rocket-lift we gave to microscopic bacteria by way of our widespread and ill-considered use of antibiotics. We successfully wiped out some generations of disease-causing bacteria, and in that process we helped nature to engineer some superbacteria, microscopic bugs undaunted by our antibiotic powers. Might we make that same mistake with insects and insecticides? It stands to reason that the insects—and the diseases they vector— can get better at doing what they do, namely biting back. So it seems to me that we should slow ourselves down, finding ways to be selective, thoughtful, even respectful in our own defense.

There is extraordinary beauty in the insect world. Picture the butterflies of spring and the dragonflies of late summer. Picture the symmetry of spider webs. I am especially fond of spiders. But I'm not so lost in silk and gossamer as to miss the point that these small animals are also dangerous, and that our human ways of being contribute not only to the evolution of superbugs but also to the ever-easier migration of insects from one continent to another, from an environment where they know and are known to a place where there are no defenses against them should they bring unfamiliar diseases with them. Just such a new insect threat was profiled in the *New York Times* in the summer of 2018: "The Asian long-horned tick, *Haemaphysalis longicornis*, is spreading rapidly along the Eastern Seaboard. It has been found in seven states and in the heavily populated suburbs of New York City."[11] The article notes that public health officials were not worried at the time of writing. "Not worried?" I hear myself repeat incredulously. I continue reading: "In East Asia, long-horned ticks do carry pathogens related to Lyme and others found in North America. But the biggest threat is a phlebovirus that causes S.F.T.S., for severe fever with thrombocytopenia syndrome. (Thrombocytopenia means abnormally low levels of platelets, which help the blood clot; a severe drop triggers internal bleeding and organ failure.)" Okay. Not worried it is; just because they can keep your blood from coagulating, no biggie.

I had been thinking about the dangers and the beauties of bugs when, just around the time that that *New York Times* article was published, I was bitten by a Mediterranean brown recluse spider that had crawled up into the sleeve of a shirt I'd left hanging on a tree limb while I worked. It was a wake-up call about the dangers of insects. Sure, I'd felt the pinch, and had taken off my shirt and shaken it out. This one just didn't seem a big deal in the context of a summer in which I had been morning and evening meal for a host of insects, mosquitoes fierce and relentless at the top of the list, ticks next—and yes, I do wear protective clothing with repellent oils. And I remove said clothing before reentering the house because I understand with stark clarity that these arachnids are disease vectors. I sometimes long to import some North American possums to feast on the ticks. I'm enchanted by the insect-guzzling geckos who monitor the walls of my home here, inside and out, but I think they could use some help in the fields. I've also been bitten by ants while pulling weeds, and had a wham of wasp stings when I mowed too close to their nest burrowed down in the soil, unbeknownst to me. Bam! Bam-bam-bam!! Yeeooowww!!

The spider got me precisely at the point of my left elbow, where the super-sensitive ulnar nerve resides. No doubt I bent my elbow that fine summer evening just as *il violino* was walking inside the fabric of my sleeve. A little over twenty-four hours later, I woke during the night to the hottest appendage I've ever had. "*Caldo, rosso, e gonfio,*" I said to the intake nurse in my first visit to the local emergency room, showing her my very hot, very red, and very swollen elbow. Charlie had bitten me (again!), and it hurt. Actually, it more than hurt. I was suffering the classic symptoms of a recluse bite, and it would take me almost eight weeks to get back to normal. I added several new words to my Italian vocabulary that night, including *febbre* (fever) and *nauseato* (yes, that). And *violino* for the spider herself, because that's what she looks like so that is how she is known locally.

The fact that I am out weeding manually almost every day instead of spraying poison means I am also spending time with bugs. It also means I begin to notice a web of relationship between the weeds and the bugs and me, though I do not yet fully understand what I notice. Why do the snails make babies inside the broken bark of trees? Why do the little black ants congregate around this particular sticky low-growing circular weed? Maybe by this time next year, I will have spent a winter month getting answers to these very local questions. Meanwhile, I see that the snails eat the fresh green leaves of my newly planted pistachio trees. I knock them off the tree with a stick, and the tree grows new leaves.

As I am knocking the baby snails to the ground, I suddenly remember being about ten years old, watching a snail and listening to a friend tell me that if you sprinkle salt on them they dissolve. Of course I had to try it. I sprinkled salt on the slug and it dissolved before my eyes: cool. But I only did it once because it was also terrible. The same friend showed me how to set fire to an ant by focusing the sun's rays on the ant through a magnifying glass. It didn't work. I begin to look like a little psychopath in my own memories. But I wasn't that. I was a kid coming into agency, as ten-year-olds do, learning how the world works and working it for all I was worth. And I was a kid doing that developmental phase in a particular time and place. I was part of that human form of extended mind that we call culture, at a time of tremendous technological progress in twentieth-century America, where nothing was as sacred as science. Science was the discipline by which technological solutions to ancient problems were allowing humans to finally conquer nature, to look to a day in the near future when our suffering would be vastly diminished, when we would live in a world clean and orderly, healthy and well fed. We revered science, the study that requires you to put aside your emotions, to focus and discipline your mind, to manipulate and measure and repeat, to predict the effects of one thing on another, like the salt on the slug. And I think now that my childhood love of scientific experiment was the way that I, a single neuron, participated in the human version of extended mind of my times, the culture of science, until such time as my individual mind was mature enough to challenge the conclusions of our extended mind.

The little scientist in me had ceased for a moment to feel for the snail and the ant, and she had felt for that moment only the pure thrill of curiosity satisfied. But I now understand—and also offer the challenge—that even that momentary setting aside of feeling for a fellow creature is really not a good thing, even in the name of science, which I have grown to love in a deeply personal way. Of science and of everything, I now say: if you're going to do something, have the guts to feel it, because that capacity for feeling evolved in animals for good reasons and it should not be unplugged. Science taught me that and so did my own gut; the two are not antithetical.

Because my farm is an organic and sustainable project, I've got an abundance of bugs whom I am not free to ignore and not free to destroy. Did I really just say "whom"? Yes, bugs are animals, too, with internal experience of the shared world out there, and bugs are essential to ecosystems, and now it's the bugs who are in danger. This is not to make any claims about insect personhood. We simply don't know enough to make any meaningful conjecture about that, in any direction, because we have never seriously asked the question. What we can say with certainty now is that the bottom of the food chain is threatening to collapse. And if we humans continue to wipe out the bottom of the food chain, it will certainly come back to bite us. So I don't get to just "off" the bugs from my world. I have to figure out how to support them without letting them destroy me or my kind.

A 2017 study based on evidence of insect populations in Germany over a twenty-seven-year period concluded starkly, "Loss of insect diversity and abundance is expected to provoke cascading effects on food webs and to jeopardize ecosystem services. . . . Our analysis estimates a seasonal decline of 76%, and mid-summer decline of 82% in flying insect biomass over the 27 years of study."[12] What we can't quite estimate yet is the effects that the collapse of the bottom of the food chain will mean for the orders of beings higher up on the pyramid. What we do know is that the food chain is one in which every participant depends on all the others for its next meal and continued existence, using a variety of strategies to get to tomorrow and on to the next millennium. "This yet unrecognized loss of insect biomass must be taken into account in evaluating declines in abundance of species depending on insects as a food source, and ecosystem functioning in the European landscape."[13] And a 2019 study describes an insect apocalypse across the United States that is driven by a fifty-times increase in the use of toxic pesticides in just two and a half decades.[14]

The loss of weeds and bugs to massive use of poisons is more complicated than we thought it would be. The weeds and the bugs go together, and the bugs we call beneficial live in a web of relationship with bugs we call noxious and dangerous. The endangered monarch butterflies need milkweed, the honeybees are dying from pesticides that were intended for creatures like saw-toothed grain beetles and for corn worm and the snails who eat my pistachio leaves. The problem is you can't target just one bug, and even if you could, that bug was having beneficial as well as harmful effects. Just think for a minute about what insects do as members of the biosphere, forming the foundation of the food web that sustains life on Earth. They provide food for amphibians, fish, birds, reptiles, and mammals. They are pollinators of many plants, but also natural controls of pest insects that feed on crops. And many insects, like the dung beetles and carpenter ants, have important roles in recycling animal wastes and dead vegetation, using the nutrients in these materials and returning them to the soil. Experts now recognize that "a diverse population of insects benefits agriculture by keeping a balance between predatory and pest insects and providing pollination services."[15]

We need to slow down and get to know the bugs and the weeds, with feeling for fellow beings as one of the tools in the research kit. It's the bugs, the bugs in all their great abundance and variety, the bugs in all their splendorous beauty, the bugs in their vigorous industry, the bugs in their ingenious creepitude, with whom we must contend. I have long been one to carry bugs outside rather than kill them, but that's a deed that is more easily done when there are one or two who survive the insecticides of urban and suburban environments than when the bugs are many and walking through the walls at night!

I am the first to admit that it's hard to feel the value of the bugs when you are chewing around the worm in your apple, or watching flies lay eggs in your ripening olives, destroying a piece of an essential crop with each egg laid. We don't feel for most bugs the same way we feel for the warm-blooded furry friends that we have in dogs and cats. But it's in exactly this context that I think it's important that increasing numbers of people are getting mystical about bees.

An estimated 70 percent of all agricultural crops are pollinated by bees: honeybees and solitary bees and bumblebees. Not only the apple trees and the grapes and, yes, the broccoli, but cool and comfortable cotton is pollinated by bees. And of course bees make honey that we humans use for food and for medicine, and propolis, also medicinal. Bees make wax that we use for candles. Bees pollinate medicinal plants and wild plants used by other species for survival. Bees are sweetness and light and healing too. And bees are animals. It's easy to forget that insects are animals, and that insects are in a vital place at the foundation of the food chain. But we all know what happens when the foundation collapses: the whole structure goes down. Bee populations, especially those of honeybees, are rapidly declining. Flowers are the food source of bees, coevolved with them, so the equation reads, no bees, no flowers. Wildflowers, bee sustenance, are weeds that we kill with herbicides.

The bees, long one of our companion species, are now in such steep population decline that they are widely considered endangered.[16] The drop in numbers of bees has been dramatic since the late 1990s, attributed largely to a new group of pesticides, the neonicotinoids (mercifully banned altogether in the European Union since 2018). The various different kinds of neonics share some features in common. They last a long time and thus accumulate in density over months and years. They are water soluble and easily move into surface waters like lakes and streams and rivers. And because they are systemic in plants, penetrating the entire structure, leaves, flowers, stems, and seeds, "they move into pollen and nectar, thereby following a direct route to exposure for pollinators."[17] Larvae exposed to sublethal doses in brood food experience reduced survival, altered metabolism, and reduced olfactory sense as adults.[18] A 2019 study concluded that the American agricultural landscape is forty-eight times more toxic to honeybees than it was a couple of decades earlier, almost entirely (92 percent) attributed to neonicotinoids, which also remain actively toxic in an environment for more than a thousand days.[19] Neonicotinoids are not thought to be directly responsible for what is called colony collapse disorder, a name given to the sudden loss of entire beehives and families, but they are thought to increase the vulnerability of bees over time by making them less efficient in sensation, foraging, and flight, and also by increasing the susceptibility of bees to natural diseases and pests. Whatever the exact mechanisms, the sudden decimation of honeybee populations is correlated with the use of the systemic neonic insecticides. And the loss of honeybees is something we seem to feel deeply.

My own desire to keep bees goes back many years to a few summer days when I assisted my beekeeping sister, Lisa McCann, with her hive inspections in California, and ever since those idyllic days with the bees I longed to establish my own hives. Then, one day here on my farm, I saw and heard the buzzing vibration of a giant paisley made of individual bees on the outside of my barn wall. I saw bees flying into holes in the stone, and I had found a few bees inside the closed building, dead on the floor, meaning that the bees were getting in through the double stone wall. This is not impossible. The bees sense the pheromones of bees gone by and exploit the existing holes to enter the space between the two layers of wall. At first I took this association with stone to mean they must be mason bees. I had heard of mason bees. I sent a photo of the swarm to my sister who told me they were honeybees and that I should call a beekeeper to catch them. Though I was too late to catch that beautiful squirming paisley that apparently had ideas of its own, it was the start of my beekeeping adventure because I then purchased a hive to be ready for the next swarm, and with that act, I was committed.

Driven by my curiosity about how and why people get mystical about bees, I started talking to beekeepers. I wanted to get a better understanding of the motivations and experiences of others who keep bees, of how they understand themselves in relation to the bees, of why they choose this work. Do they think they are saving the bees? If so, from what and for what? And I learned this: every one of them described being stung, learning to pay attention, and delightedly discovering ways in which the bees are their teachers, their superiors.

One of them actually went to the bees looking for stings in the first instance. My sister Lisa became interested in honeybees when she learned that bee venom was a recommended alternative treatment for joint pain. She had received a diagnosis of advanced Lyme disease and hoped to avoid some of the toxic effects of standard medical treatment, such as long-term use of antibiotic cocktails. She began to study bees and soon installed three hives in her back garden. Lisa ultimately went down the road of mainstream treatments for the disease and, after some years of illness, she wasn't able to care for her hives in the way that she had planned. But she came to see herself in the bees and the bees in herself, all together struggling to survive new threats. More than ten years later, her bee families were still living in the hives: they were surviving, perhaps even thriving. They may have swarmed when she was not paying attention, yet new generations remain.

Lisa reports feeling deep satisfaction now when she observes the scouts returning from explorations, making their signature figure-eight waggle-dance, telling other bees where to go to find food.[20] This is another way that bees are like humans, in their use of language as one way to be in constant communication with each other; linguists recognize the waggle-dance as true language. Lisa explains that the fact the bees are doing the waggle-dance, using their language, is an indication that her hives are home for these bees, that they are thriving families, and not just a collection of robbers looking for old honey in the frames. "I think they have been fighting to survive all along, as hard as I have," she says. "That's the hardest part to express . . . after all the scientific observation, experimenting, note-making, analysis, trials of this and that, how do you express the knowledge of resonance? And the understanding that what connects it all is spirit? The knowing that you are connected by more than a fight to live?"

Lisa's description of the changed relationship she has to the bees in her garden reminds me of the descriptions of some human relationships: with pressures eased, the pleasure of being together increases. But I wonder whether the bees noticed that Lisa had to leave them for a while because of her own ill health, and that she has now returned? Do they, too, say, "Ah, that"? Do they know that they are cohabiters of an ecological niche, with at least one cohabiter consciously in relationship? There is a strong strand of bee lore, historic and contemporary, that says the bees do understand us. Is it fanciful folklore, or might there be some objective truth in the claim? We don't know. Yet.

Tabita Faoro—an Italian woman living in France who began learning about bees when her philosophy professor illustrated his lectures with experiences drawn from his beehives—expressed to me an equally strong sense of connection to the bees for whom she cares. "Bees are my practice. They teach me how the natural laws of the cosmos work. They have hyperdeveloped auditory and perceptive systems, and they perceive the intentions of those approaching. They do not tolerate brusque gestures and nervousness and fear. In short, you can never lie to a swarm!"

The bees wake you up. The bees keep you honest. The bees perceive you, accurately. And the bees somehow teach you to understand them. Tabita added that she developed with her bees something that she experienced as intuition. "Several times I wanted to visit the hive because I felt something had changed, and I was never wrong. There is something mysterious and inexplicable in the relationship that is established, something that opens the doors of eternity. If we think that a bee lives forty days—it's not like the relationship that you have with a kitten, which fills our little affective voids—the relationship with the bees remains in the sphere of the mind. At the same time bees are joy, vitality, they are the life that is generated and regenerated but is always the same and always renewed."

And like every beekeeper I talked with, Tabita sees the bees as being superior in some ways, a kind of model for human behavior. "The bees know how to take care of themselves very well, but we begin through them to take care of ourselves: to have the courage, each of us, to realize our own duty, according to our own nature, to assume our own responsibility to be an integrated part of our own species, like the bees are in theirs, to do everything possible to put ourselves at the service of the collective good."

Decades after my first bee-rescue misadventure, I finally became a beekeeper in 2019. I took the basic beekeeping certification course in my new language, Italian, counting on the fact that I'd already acquired a fair amount of bee knowledge to get me through. Through that course, I also acquired a dedicated pair of bee coaches, Giovanni Guarini and Roberto Caforio, who brought to me one of my two hives, the blue hive with the metal roof and runway. The other hive, the gold one, we assembled together, the wood flowhive that I'd purchased in the hope of being ready for the next swarm. Days later, Giovanni and Roberto captured a swarm in the lemon grove, and we installed them in the gold hive.

Giovanni and Roberto are entering their fourth year of beekeeping and they manage more than thirty hives together. Exiting from the worlds of business and finance, they have made a life change to advocacy for bees, political action for the environment, and sustainable living for themselves. They hope to make of their beekeeping a means of supporting themselves, but they make abundantly clear that the health of the hives is their first priority and that any profit from the sale of honey is the last. For them, keeping bees is a way to be connected to nature and to live sustainably, and to teach others how to live sustainably by telling the story of their honey when they sell it.

Giovanni unapologetically refers to the bees as superior to humans, having something to offer that, right now, we need. "Their inspiration for us is the way they work together, their dedication to their work, their industriousness," he says. "We have to save the bees because they know how to live! They give us a model of how to live, a perfect society. Everyone is equal (except the queen), everybody works for the same goal, which is to make their species survive in the ecosystem. Our species, we perpetuate ourselves by exploiting resources. Bees perpetuate themselves by enriching the environment, not exploiting it: bringing more biodiversity, pollinating, giving services to the ecosystem." Giovanni sees in bees a more cooperative way of being, a kind of ecosystemic orientation that we humans lack, but that we might learn from the honeybees.

Tabita Faoro also thinks about the importance of valuing the bees for themselves and not as resources to be exploited. "A teaspoon of honey is what a bee produces in forty days of work, her whole life, and we eat it on bread without thinking about it. I think that we have lost the value of the sacred, of the medicine that the sacred represents, of magic. But as long as there are bees, neither the sacred, nor the medicine, nor the magic is lost to humanity." Tabita sees the bees as teaching her a kind of wakefulness, attentiveness, and also the capacity to regulate her own emotions as a prerequisite to being in relationship with them. Describing her first solo forays in doing hive inspections, she says, "When, months later, I found myself making visits alone, I had to step back and take a normal breath before making a second attempt, because the bees attacked me just as I opened the hive." Slowly, she learned how to approach them without anxiety, without an attitude of preoccupation. "There are ways of acting, ways of smelling, ways of appearing that the bees don't like, and they let you know they don't like these things by stinging you."

I can attest to the fact that, even without any drama attached to it, several bee stings at one time is a very uncomfortable condition. Until I had several stings at once on my hands, I preferred to work the hives without gloves, and I still prefer to work gloveless with swarms because swarms are a collection of well-fed, calm, and generally docile bees. Without gloves I am better able to feel them, less apt to squish them, more sensitive to the collective mood. Swarming bees are not looking for trouble; they are looking for a new home. But the fact is that when you have been stung, you are marked by scent as a problem person until the pheromones dissipate, and you will attract more stings because of it. The bees have chastised me, and now I understand that there is a time and place for the preventive power of protective cover from stings. Tabita told me that her own mentor asked her to work barehanded, except in the case of prior stings. "Pierre allowed me to wear gloves only in one case: when they had already stung me, due to imprudent or too fast movements. The sting secretes an unpleasant odor and communicates to the other bees that this person who has been stung is an enemy, thus increasing the possibility of further stings. But bees do not sting if they don't feel attacked, and if they do it is always to protect the family." The bees, somehow, are never at fault. Rather, all the beekeepers blame themselves (or each other) for not correctly understanding, or for behaving in bee-offensive ways.

Giovanni and Roberto started beekeeping when an older man, a bee-keeper in a nearby town, wanted to retire and sell everything. They decided to buy the hives and equipment, and to acquire the education they would need to manage the bees. Giovanni says that it is important to get some formal education about bees, but that the bees themselves teach you what you most need to learn. "It's good to have education about bees, because you need to know about the cycles, about their way of life, their organizing." But for him, too, the knowledge-building is a small part of beekeeping because "it's only your experience that allows you to enter into relation with them. Besides all the techniques, the big thing is to enter into relationship with the bees and this is what teaches you." Roberto adds, "Everything for us started with curiosity, but this curiosity became a passion for us, the will to discover more about the life of the bees, their social life, the structure of the family. It went from curiosity to passion. Now bee life is a metaphor for our life: if we save bees, even if we do it egotistically or selfishly, we save ourselves too."

Like Tabita, Roberto thinks that the bees taught him a different—and better—way of being. "In my case, the relationship with bees and beekeeping, it was really intense because it developed my powers of observation, my spirit of observing other animals approaching bees. I began observing the movements of bees, the movements of other insects, observing the flowering of the plants. It's also the sensation that I have when I open the hive, it's like being in a parallel world. The way that you observe things changes. You're in the present, it's much more concentrated. It's a kind of entering into this world and isolating yourself from everything else. You can be there from the morning until one in the afternoon, and you don't feel time passing. Five hours as if it were a few minutes. And all your problems are gone. It's you, the bees, and that's it."

The fact that the bees wake you up is a common motif, and the moral of that story is in the positive experience of being in the flow, where you lose track of time and are completely absorbed in the tasks at hand. But it is also seen in the negative experience of being punished by the bees if you fail to find your center, if you bring restlessness, fear, turmoil with you. "But also it happens that when you enter this world, with your problems and your thoughts, if you're nervous and you're thinking about something else, the bees notice it and they start moving in a different way. So you have to learn that when you approach them, you need to be focused," Roberto says. And I hear the unspoken refrain: "Or you will get stung."

"At the beginning, we weren't aware," Giovanni chimes in, "and maybe we were moving more like an elephant in a crystal shop. The bees also know us better now, and it's this process of knowing each other, and knowing how to move with respect to the bees, how to cooperate. This approach is essential. If you're thinking about something else, they sense it."

Clearly, these two men have invested a great deal of themselves in learning how to be with the bees, in acquiring education, through both books and the corrective stings of their bee mentors. Like other animal rescuers I have met and interviewed, these two don't see an end in sight, a time for giving it up and doing something else. "It's not always easy, especially when you are starting up, as we are," explains Giovanni. "The problem is that you have to deal with an animal, with an insect, not an object. So when you create this relationship, it's like saying goodbye to a person if you were to quit. No! It's not possible." For Roberto, too, the bees allow no escape, and he wouldn't want it to be otherwise. "From my point of view, when you're in, you're in. Maybe from an economic perspective, you might not have the resources . . . or you need more resources to have your own hives, or you need more revenues. But quitting is impossible. You can read more, you can find more information, but even if you don't have bees any more, you can't ever get out of this mindset. You can't quit! Your mind is always there. You can quit with your body but never with your soul."

Such different entrances into the world of beekeeping: seeking a treatment for joint pain, approaching bees as a philosopher looking for alternative social organizations, and wanting to work for political change in relation to the environmental crisis and at the same time to make a honey business integrated into a rural homestead. Yet through these various motivational portals each person came to see him- or herself as being in a personally transformative relationship with the bees. Even the stings bore witness to this relationship, for the bees sting when the humans are out of touch, out of harmony with the bees. To what other creature do we attribute such superiority that when they hurt us, we assume it was to correct what we had gotten wrong? Yet every beekeeper to whom I have ever spoken has attributed superiority to the bees, including saying that stings are wake-up calls to something that we humans are doing wrong.

How are we all doing this, I wonder, putting up with itching bee stings, and with the fear of them, and still giving all the credit to the bees for good, and all the credit for malice and ignorance to our own human selves? Are we collectively projecting some mystical power onto the bees in hopes perhaps of incorporating that power into our own selves? "How many times have we been stung? We don't know, we lost count! You can go from zero stings in a month to fifteen in a day. It depends. Maybe you make a mistake, or you're not in the mood to work with the hive, and then maybe they are more aggressive." Roberto laughs and adds, "The intriguing thing with being stung is that they can sting you in places that you could never even think about it! In parts of the body that you don't even imagine! The ankle, the centimeter of skin that wasn't covered, when they go through the mask with the sting and sting you on the nose! The one on the nose is the worst! For half an hour, you cry."

If my contemporary beekeeping acquaintances are strangely mystical about their relationships with the bees, at least it is nothing new. The Egyptian great mother goddess Neith was worshiped at Sais and her temple was called the House of the Bee. Minoan, Mycenaean, and Greek ancient cultures all contained bee cults, associated with gods, goddesses, and the *melissae*, priestesses of the bee cult. The Celts had an esoteric bee cult and practiced a kind of bee-centered shamanism. "The narratives from antiquity," Michaela Fenske says, "are extremely positive: the praise of bees was a specific genre in (popular) literature and poetry. There is hardly any other insect praised so much because of its virtues."[21] And a contemporary project in sound, *Hymns from the Hive*, celebrates the particular sacred relationship of women and bees.[22] Perhaps we have long revered the bees because we sensed in them a different kind of wisdom, and a kind of wisdom that we especially now need to learn.

In 1858, John Greenleaf Whittier wrote a poem called "Telling the Bees," in which he intentionally immortalized the tradition of informing the bees about important events in the lives of humans, especially deaths in the families of beekeepers. As Colleen English notes, "The emphasis that Whittier places on this concept of delivering important information to the bees implies that there is a special relationship that exists between honeybees and humans that is essential to maintain."[23] New research has discovered that an old relationship between African honey-hunting humans and their honey-guide birds is even deeper than we knew. The birds understand human speech, and this "suggests that the honeyguide and human behavior have coevolved in response to each other."[24] Honeyguide birds seem to be yet another species that coevolved with humans, raising the question of whether we are not all coevolved with one another in a great web of relationship, affirming both the scientific description of evolutionary continuity and the age-old human mystical leaning toward identifying with the whole of life, and leaving us holding that bag full of the question about how we are to live if all are somehow one.

Meanwhile, though, it seems as though there is indeed something archetypal in the relationship of humans to bees, a kind of expectation that bees have some special capacity to introduce us into the secret operations of the natural world. Michaela Fenske suggests that bee swarms, viewed in the eighteenth century as symbolic of the chaos of nature, are now imbued with symbolism of a kind of creative democratic spirit that might remake humanity in a more ecologically viable form.[25] Some people think that we could solve the problem of ecological collapse by moving to Mars, but I think that our swarm, if it happens, will be to find a new way of living, even a new way of moving from one way of being into another, rather than expanding our geographical frontiers one more time. Perhaps we will learn to use our extended mind and collectivity in more cooperative ways. We might need not a new place to live, but a new way to live in this same ever-changing place that is Earth. Of this need to find a new way, Fenske offers the image of the swarm: "Bee swarming has increasingly become a positive sign for an independent, powerful nature."[26] If we can learn to swarm, we can perhaps return home to nature again, not as masters, but as members.

I will never forget the capture of my first swarm, a tiny group of a few hundred bees who were weaving wax combs onto a few leaves of one of my lemon trees. We had brought the hive to where they were, installing it on a bench made of boards and bricks. Giovanni instructed me to take the roof off the hive and then the top board too. We each held onto one of the main tree branches on which the bees were making new comb, and Roberto cut the branches just outside our hands. When the lemon branches were freed from the tree, complete with the bee swarm, we located the queen of the small family and placed her gently in a small cage designed for this purpose, to be released on the following day after all the other bees had followed her scent. Giovanni instructed me to put the little cage in the bottom of the hive, and I did so. Next he told me to take the branches containing the swarm and shake them over the open hive. I did, and just like that, down went more than half of the bees to join their queen. Now I covered the hive box with the top board, its large circular hole in the middle. And here began the most extraordinary show, with first one bee, then two, then three standing on the edge of that circle, each raising her rear to release the pheromones that say this is home, and then using her wings to fan those pheromones into the air, attracting more bees. Soon there was a circle of bees pumping pheromones into the air and fanning with their wings, and then that circle became a spiral of bees pumping pheromones and fanning them in the air, and then the spiral began to descend down into the hive. How can we begin to understand this, a thousand individuals appearing to reach perfect consensus and cooperation in declaring a place home, and dedicating themselves to that home?

I recall times in which insects were derided for not having minds of their own, when this case was cited as illustrative of human superiority: we with our individual minds can master individual learning and creativity, and we can make individual choices, or so we bragged. But now *hive mind* is the term we use to indicate our capacity to take advantage of collective knowledge. And "hive mind," we increasingly understand, describes our own extended and distributed mind, that thing we call culture that inhabits us as individuals and shapes us, even as individuals then reshape it.

Says Giovanni, "You see that in a sense there are no choices, because the planet is giving us an expiration date, which is actually very near. If we want to live in this beauty that is Earth's gift to us, and if we want to commit, even if society is not ready, we have to start somewhere. Someone has to start spreading the message." I think that Giovanni and Roberto intend that their individuality will influence their human hive-family to move, not to a different location, but to a different way of being human, a more hive-like way in which each offers himself for the good of the whole. As Fenske puts it, "Today's stories about swarming bees have also become stories about alternative ways of life and new beginnings."[27]

I picture Giovanni, Roberto, Tabita, Lisa, and me in our local places, singular, each of us, sending out the pheromones to tell the neighbors that this place is home, fanning those pheromones with all our might, knowing that the swarm comes home by agreeing with that first bee's message that this is home. So begins what may yet become a great spiral of mutuality that allows for our lives to go on together. This, I think, is the hope of the new breed of beekeepers, that nature will not be mastered by humans, but inhabited by us, mindfully in relationship to all the others.

The bee way gives us another approach to the experience of biophilia, the term famously coined by Edward O. Wilson to indicate that we humans—and probably all life-forms in their various ways and according to their various genetic endowments—naturally feel love for the web of life. Biophilia, I might suggest, is very close to that most fundamental affect of mammals, the neural seeking system. The seeking system is part of the wiring inside of each animal that wakes it to the feeling for being in relationship with a world out there, to which it belongs. It is perhaps the most fundamental feeling in living, to seek to encounter the other, and to live in relationship.

At the moment when the Covid-19 pandemic was first sweeping Italy, I learned that my nephew Joe had died in California suddenly and unexpectedly. The brakes had been thrown on, and everything was on sudden hold in my life. I was in quarantine while my family was in crisis far away. I woke early one spring morning, a few days after Joe's death, in self-isolation on the farm, to the sound of my brain's declaration, "The bees need water in their bowl!" Compelled to care for them, I found that indeed there were a few bees trying to suck water from the drying muck in the bottom of their bowl. In my fervor to help, I almost accidentally drowned them with the hose. Seeing this, I let my eyes open to them, felt my own heart settle into itself, and then I gently and barehandedly lifted the bees and set them on higher ground. I softened the flow of water into a nice swirl as I filled their bowl. Then I sat on the bench next to them and sang the song I always sing, a lullaby, the idea being that I will create for them a positive association among the variables: my voice and pheromones, their extra food rations and water. The hive responds to me with a kind of gentle curiosity. I suddenly remembered that I am supposed to tell them about important events in the family. I told them Joe had died, I told them about the Covid-19 pandemic that was attacking my kind.

As I sat with them, I watched the magpies swooping and lunging for worms in the newly plowed fields, and I remembered my raven friend Jason, a continent away in space, a decade past in time, Jason with his amazing raven brain, that compact and well-organized tiny organ that provided him with so much of what I have in my big primate brain, his language so much like mine that we could play games of syntax together. And now I think of the bees, with their distributed intelligence, each bee being something like the equivalent of a neuron, and all of them together sensing, interpreting, deciding in their democratic way, using their language of dance. We are so full of our individual selves and the capacities of our individual minds. And they seem so selfless, so altruistic. But I wonder if it's even true that we, the deeply enculturated humans, are as separate and independent as we habitually think we are? Or that they, each but a neuron in a hive mind, are as lacking selfhood as we think they must be? Why would they need influencers if they were hardwired to be group and only group? We don't know, but the strength of the dance inclines me to think that we share some powerful similarities.

The other may feel close to us, as our families do, as our dogs and cats do, and as virtually all mother and baby mammals do, even the huge whales.

The other may feel more distant, as a spider or a beehive does. But the others all must somehow be welcomed if we are to swarm to a new way of being in which we are part of them and they are part of us, all together holding so many places of service in a new behavioral field of being. Can we humans actually do this? Intentionally move ourselves out from our so recent perception of ourselves as masters of the universe, to roles of service within a field of being that even includes bugs? Some, at least, of our human imaginal discs are cultural, and some of them are imagining that we might learn this kind of social transformation from bees.

It gives me hope that we might yet swarm.

NOTES

Chapter 1. In Times of Drought

1. Sewall, "Girl Who Gets Gifts."
2. Linz, Knittle, and Johnson, *Ecology of Corvids.*
3. Olkowicz et al., "Birds Have Primate-Like Numbers."
4. Heinrich, *Life Everlasting*, 69–70.
5. Ibid., 61.
6. Marzuff and Angell, *In the Company of Crows and Ravens.*
7. Benvenuti, *Spirit Unleashed*, 58.
8. Boxall, "Another Toll of the Drought."
9. Asner et al., "Progressive Forest Canopy Water Loss."
10. Oerlemans, *Living Planet Report*, 18.
11. Ibid., 44.
12. Kolbert, *Sixth Extinction.*
13. La Ganga, "Prisoners Rescue Deaf Dogs."
14. Benvenuti, *Spirit Unleashed*; Benvenuti, "Evolutionary Continuity and Personhood"; Benvenuti, "Evolutionary Continuity of Personhood."
15. Morell, *Animal Wise*, 93–115.
16. Call and Tomasello, "Does the Chimpanzee Have a Theory of Mind?"
17. Johnson, *Ask the Beasts.*
18. Goodall and Bekoff, *Ten Trusts*, xv.

Chapter 2. Hard Lives, Soft Landings

1. Mercy for Animals, *Farm to Fridge.*
2. For an exposition of this, see Balcombe, *What a Fish Knows.*
3. Donaldson and Kymlicka, *Zoopolis.*
4. Benvenuti, "Evolutionary Continuity and Personhood."
5. Cunniff and Kramer, "Developments in Animal Law."

6. Donaldson and Kymlicka, "Farmed Animal Sanctuaries."
7. Ibid., 50.
8. Walker, *Chicken Chronicles*, 13, 15.
9. Akhtar, *Our Symphony with Animals*, 192.
10. Farm Sanctuary, *Julia: An Incredible Pig*.
11. Panksepp and Biven, *Archaeology of Mind*.
12. Ibid., 17.
13. Ibid., 313.
14. Panksepp and Biven, *Archaeology of Mind*; Damasio, *Strange Order of Things*; Cacioppo and Patrick, *Loneliness*.
15. Panksepp and Biven, *Archaeology of Mind*, 318.
16. Watson and Watson, *Psychological Care*, 81–82.
17. Bowlby, *Childcare*; Spitz and Wolf, "Anaclitic Depression"; Harlow, "Nature of Love."
18. Panksepp and Biven, *Archaeology of Mind*, 321.
19. Goodall and Bekoff, *Ten Trusts*, as quoted in chapter 1, n16.

Chapter 3. The Big Five and the Little Five Million

1. El Jundi et al., "Straight-Line Orientation."
2. Dacke et al., "Dung Beetles," 298.
3. Heinrich, *Life Everlasting*, 145.
4. Ibid., 133.
5. Hoedspruit Endangered Species Center, "HESC Founder Lente Roode."
6. Adams, "Bush Baby."
7. National Geographic Society, "Black Mamba."
8. Panksepp and Biven, *Archaeology of Mind*, 2–3.
9. Animal Adventure Park, "April the Giraffe."
10. Panksepp and Biven, *Archaeology of Mind*, 265.
11. Cristina Eisenberg, foreword to Colinvaux, *Why Big Fierce Animals Are Rare*, xi–xii.

Chapter 4. It Starts with One

1. O'Connell, "Secret Lives of Elephants."
2. O'Connell, *Elephant's Secret Sense*.
3. Dell'Amore, "Elephants Have 2000 Genes for Smell."
4. Panksepp and Biven, *Archaeology of Mind*, 351–88.
5. Ibid., 357.
6. Panksepp, "Rats Laugh When You Tickle Them."
7. See Damasio, *Strange Order of Things*; and Panksepp and Biven, *Archaeology of Mind*, for in-depth discussions of affect as evolved expression of biological values.
8. Panksepp and Biven, *Archaeology of Mind*, 382–84.
9. Bekoff, "Animal Emotions," 867.
10. Panksepp and Biven, *Archaeology of Mind*, 422.
11. Christy, "How Killing Elephants Finances Terror."
12. Camp Jabulani, "Our Elephant Herd Instrumental."
13. Message from Adine Roode, managing owner, in Camp Jabulani, *Camp Jabulani*, 1; ellipsis in original.

14. Ram, "Power of One."

15. Singer, *Animal Liberation.*

16. Mill, "Coleridge," 220.

17. Had we counted insects among the total number of beings, we would have done everything to promote the well-being of insects, and not just focused on humans, and human subgroups at that.

18. See Damasio, *Strange Order of Things*, for a full exposition.

Chapter 5. Endangered

1. Frost, "Mending Wall," in *Collected Poems*, 33.

2. Hoedspruit, "Cheetah Breeding Project."

3. Ibid.

4. CNN Wire, "Formula E Car vs. Cheetah."

5. See, for example, Slobodchikoff, *Chasing Doctor Dolittle.*

6. Darwin, *Descent of Man*, 173.

7. I explore this in more detail in *Spirit Unleashed*, 13–17.

8. Riddle, "How Bees See."

9. Shubin, *Your Inner Fish*, 178.

10. Benvenuti, "Evolutionary Continuity of Personhood"; Rowlands, *Can Animals Be Persons?*

11. You can see photos of veterinarians providing Phillipa's treatment on the HESC website: http://hesc.co.za/2016/06/philippas-treatment-update-11-june.

12. Hoedspruit Endangered Species Centre, "Ike, a Strong Rhino Bull."

13. Social neuroscience does suggest hypotheses regarding mirror neurons and muscle mimicry. See Churchland, *Braintrust*; Rizzolatti, Fabbri-Destro, and Cattaneo, "Mirror Neurons."

14. Panksepp and Biven, *Archaeology of Mind*, 419.

15. Benvenuti, *Spirit Unleashed*, 145.

16. I explore this in documented detail in chapter 6.

17. London, *White Fang*, as quoted in Panksepp and Biven, *Archaeology of Mind*, 175; ellipses in Panksepp and Biven.

18. Panksepp and Biven, *Archaeology of Mind*, 176.

19. Levine, *Waking the Tiger.*

20. Darwin, *Descent of Man*, 45.

21. Gilford, "Why Does a Rhino Horn Cost $300,000?"

22. Mogomotsi and Madigele, "Live By the Gun."

23. Carnie, "Understanding Demand for Rhino Horn."

24. Rhino horn consists of keratin rather than dentine, and therefore from the scientific point of view isn't technically ivory. However, the term "ivory trade" commonly includes rhino horn, and I use "ivory" in that sense.

Chapter 6. Oxytocin Bay

1. Scammon, *Marine Mammals*, 29. Scammon captained many whaling expeditions but later became a champion of the whales.

2. Hof and Van Der Gucht, "Cerebral Cortex of the Humpback Whale," 1.

3. Payne, *Among Whales*, 21.

4. As referenced by McQuay and Joyce, "It Took a Musician's Ear."
5. Fitch, "Evolution of Language," 199.
6. Marino et al., "Cetaceans Have Complex Brains."
7. Fields, "Are Whales Smarter Than We Are?"
8. I tell this story in *Spirit Unleashed*, 131–36.
9. Editorial Multimedia, "Laguna San Ignacio."
10. Aridjis, "Savior of the Whales."
11. Aridjis, "The Eye of the Whale," in Russell, *Eye of the Whale*, 14.
12. Siebert, "Watching Whales Watching Us."
13. Ibid.
14. Schore, "Effects of a Secure Attachment Relationship," 7.
15. Ibid.

Chapter 7. Why Cats, Florence?

1. Pierce, "Why Do Cats Meow at Humans?"
2. Panksepp and Biven, *Archaeology of Mind*, 97–144.
3. Ibid., 98.
4. Bradshaw, *Cat Sense*, 11–34, especially 31.
5. Ibid., 7.
6. Ibid., 37.
7. Thomas, *Tribe of Tiger*, 173.
8. Nightingale, *Notes on Nursing*, 58.
9. Thomas, *Tribe of Tiger*, 73.
10. Dowling, "Complicated Truth about a Cat's Purr."
11. Lyons, "Why Do Cats Purr?"
12. Dowling, "Complicated Truth about a Cat's Purr."
13. Bradshaw, *Cat Sense*, 7.
14. Kotrschal et al., "Human and Cat Personalities," 115.
15. Ibid.
16. Wedl et al., "Interactions between Domestic Cats," 58.
17. Bradshaw, *Cat Sense*, 59.
18. Vitale, Behnke, and Udell, "Attachment Bonds between Domestic Cats," 865.
19. Bradshaw, *Cat Sense*, 1.
20. Berkeley, *TNR, Past, Present, and Future*.
21. Catnippers, *Overview and Key Facts*.
22. Ibid.
23. Loss, Will, and Marra, "Impact of Free-Ranging Domestic Cats."
24. Royal Society for the Protection of Birds, "Are Cats Causing Bird Declines?"
25. Ibid.
26. Thomas, *Tribe of Tiger*, 260.

Chapter 8. Love Dogs

1. Cacioppo and Patrick, *Loneliness*, 257.
2. Berns, Brooks, and Spivak, "Canine Brain Responses."
3. Clive Wynne, *Dog is Love*.
4. Wynne, quoted in Karin Brulliard, "What Makes Dogs So Special."

5. Senese et al., "Human Infant Faces."
6. Rumi, "Love Dogs," in *The Essential Rumi*, 155.
7. Hobgood-Oster, *Dog's History of the World*, 17.
8. Ibid., 10.
9. Panksepp and Biven, *Archaeology of Mind*, 154–56.
10. Ibid., 147.
11. Siegel, *Neurobiology of Aggression and Rage.*
12. Panksepp and Biven, *Archaeology of Mind*, 153.
13. Chanowitz and Langer, "Premature Cognitive Commitment."
14. Rizzolatti and Sinigaglia, *Mirrors in the Brain.*
15. Churchland, *Braintrust*, 148.
16. Overman, "What is Disorganized Attachment?"
17. Potter and Mills, "Domestic Cats (*Felis silvestris catus*)."
18. Thank you, forever, to Mary Oliver for giving us such a poignant poetic image in the poem "Wild Geese."
19. Rich, "Integrity."

Chapter 9. Animal Ambassadors

1. Holdgate, *John Herbert Strutt*, 7.
2. Pepperberg, *Alex and Me.*
3. Holdgate, *John Herbert Strutt*, 5.
4. Ibid., 2.
5. Schwarz, "Free-Range Parrots."
6. The reference is to Judith Viorst's *Alexander and the Terrible, Horrible, No Good, Very Bad Day* (1972), a classic children's book and favorite cultural reference.
7. Monkey World Ape Rescue Centre, orientation brochure, undated.
8. Ibid.
9. Saint-Exupery, *Little Prince*, 33.
10. Kemmerer, "Nooz: Ending Zoo Exploitation," 37–38.
11. Smithsonian National Zoological Park, "Shanthi, the National Zoo's Musical Elephant."
12. Thomas, *Tribe of Tiger*, 215–29.
13. Ibid., 214.
14. Ibid.
15. Weiss et al., "Evidence for a Midlife Crisis."
16. Gilmour, "'Fleas Were Eating Animals Alive.'"
17. Gooden, "California Living Museum Accused."
18. Nicolas Brulliard, "Wild Beasts of the Urban Jungle."
19. Benson et al., "Extinction Vortex Dynamics."
20. Sanchez, "California to Build."
21. Ibid.
22. Tucker et al., "Moving in the Anthropocene."
23. Wilson, *Half-Earth.*
24. Solly, "California Will Build."
25. Harding, *Rabbit Effect*, xxiii.
26. Earthfire Institute, "Dr. Con Slobodchikoff," 9.
27. Public Broadcasting Service, *Koko: The Gorilla Who Talks.*

28. Main, "Why Koko the Gorilla Mattered." I note the discrepancy in dates between the PBS documentary, which features Patterson and gives Koko's date of birth and the beginning of Patterson's research as 1971, and the *National Geographic* article, which gives the research start date of 1972.

29. Koko lived with two males, Michael and Ndume, for most of her life.

30. McGraw, "Koko Mourns Kitten's Death."

Chapter 10. The Greatest Story on Earth

1. Oerlemans, *Living Planet Report*; Kolbert, *Sixth Extinction*.

2. Bulman, "Villagers Knit Jumpers for Indian Elephants."

3. Slobodchikoff, *Chasing Dr. Dolittle*, 28.

4. Ibid., 27.

5. Low, "Cambridge Declaration on Consciousness," 2.

6. Griffin, *Question of Animal Awareness*.

7. Benvenuti, *Evolutionary Continuity of Personhood*, 3.

8. Ager, "Ringling Will Retire Circus Elephants."

9. Actman, "For Dolphins, a Bold Decision."

10. Karin Brulliard, "One Problem with Shutting Down the Circus."

11. Shamoo and Resnik, *Responsible Conduct of Research*.

12. CBS News, "Animal Crackers' Animals 'Freed'."

13. "Declaration of Rights for Cetaceans."

14. Benvenuti, "Evolutionary Continuity of Personhood."

15. Rowlands, "Are Animals Persons?"

16. Schneider and Kay, "Tilikum, Orca That Killed Trainer, Dies."

17. Bonal, "Circular: Policy on Establishment of Dolphinarium."

18. Winter, "Court Declares Captive Orangutan."

19. Saucier-Bouffard, "Legal Rights of Great Apes," 235.

20. Cunniff and Kramer, "Developments in Animal Law."

21. Donaldson and Kymlicka, *Zoopolis*, 50–68.

22. Associated Press, "California Becomes First U.S. State."

23. Sullivan, "Historic Victory!"

24. Canon, "'A Loud and Clear Message.'"

25. Corcoran, "California's Economy."

26. The painting is in the collection of New York's Metropolitan Museum of Art and can be seen at https://www.metmuseum.org/art/collection/search/484972.

27. Baird, *White as the Waves*; Gowdy, *White Bone*.

28. Paull, *The Bees*.

29. Kingsolver, *Prodigal Summer*.

30. Fowler, *We Are All Completely Beside Ourselves*; Doyle, *The Plover*; Doyle, *Martin Marten*.

31. Peterson, "Religious Studies and the Animal Turn," 233.

32. Ibid.

33. Ibid., 238.

34. Slobodchikoff, *Chasing Dr. Dolittle*; Cornell Lab of Ornithology, Center for Conservation Bioacoustics.

35. Mychalejko, "Ecuador's Constitution Gives Rights to Nature."

36. Blackiston, Casey, and Weiss, "Retention of Memory through Metamorphosis."

37. Orr, "Tale of a Constipated Goldfish."

Chapter 11. Might We Yet Swarm?

1. Zimmer, "To Bee."
2. Tautz, *Buzz about Bees*, prolog.
3. Hölldobler and Wilson, *The Ants*; Hölldobler and Wilson, *Superorganism*.
4. Cacioppo and Patrick, *Loneliness*.
5. Cacioppo, "Lethality of Loneliness."
6. Winnicott, *Child, Family, and Outside World*, 88.
7. Pert, *Molecules of Emotion*.
8. Cornell, "Honeybee Decision-making Ability."
9. Seeley, Camazine, and Sneyd, "Collective Decision-Making in Honey Bees," 286.
10. Fenske, "Narrating the Swarm," 131.
11. McNeil, "Invasive New Tick."
12. Hallmann et al., "Decline in Total Flying Insect Biomass."
13. Ibid.
14. Leahy, "Insect 'Apocalypse' in the U.S."
15. Lundgren and Fausti, "Trading Biodiversity for Pest Problems."
16. Tirado, Simon, and Johnston, *Bees in Decline*.
17. Code, "Neonicotinoids," 17.
18. Hopwood et al., *Recommendations to Protect Pollinators*.
19. DiBartolomeis et al., "Assessment of Acute Insecticide Toxicity."
20. Attenborough, "What's the Waggle Dance?"
21. Fenske, "Narrating the Swarm," 134.
22. Redmond and Mascarenhas, *Hymns from the Hive*.
23. English, "Telling the Bees."
24. Spottiswoode and Begg, "Reciprocal Signaling."
25. Fenske, "Narrating the Swarm."
26. Ibid., 132.
27. Ibid., 133.

BIBLIOGRAPHY

Actman, Jani. "For Dolphins, a Bold Decision by the National Aquarium." *National Geographic*. June 15, 2016. http://news.nationalgeographic.com/2016/06 /national-aquarium-captive-dolphins-retire-ocean-sanctuary.

Adams, Susan. "Bush Baby." *Forbes*. March 29, 2004. https://www.forbes.com/global /2004/0329/060.html#2b2560c51f90.

Ager, Susan. "Ringling Will Retire Circus Elephants Two Years Earlier than Planned." *National Geographic*. January 11, 2016. http://news.nationalgeographiccom/2016/01 /160111-ringling-elephants-retire.

Akhtar, Aysha. *Our Symphony with Animals: On Health, Empathy, and Our Shared Destinies*. New York: Pegasus, 2019.

Animal Adventure Park. "Animal Adventure Park's April the Giraffe—Live Birth." April 14, 2017. YouTube video. https://www.youtube.com/watch?v=dz_errIBq3Y.

Aridjis, Homero. "The Eye of the Whale." In Dick Russell, *Eye of the Whale: Epic Passage from Baja to Siberia*, 12, 14. New York: Simon and Schuster, 2001.

———. "Savior of the Whales." *New York Times*, October 31, 2013. https://www.nytimes .com/2013/11/01/opinion/savior-of-the-whales.html.

Asner, Gregory P., Philip G. Brodrick, Christopher B. Anderson, Nicholas Vaughn, David E. Knapp, and Roberta E. Martin. "Progressive Forest Canopy Water Loss during the 2012–2015 California Drought." *Proceedings of the National Academy of Sciences* 113, no. 2 (2016): E249–E55. http://www.pnas.org/content/113/2/E249 .abstract.

Associated Press. "California Becomes First US State to Ban Animal Fur Products." *Guardian*, October 12, 2019. https://www.theguardian.com/world/2019/oct/13/fur -ban-california-outlaws-making-and-selling-new-products.

Attenborough, David. "What's the Waggle Dance? And Why Do Honeybees Do It?" Smithsonian Channel. May 20, 2016. YouTube video. https://www.youtube.com /watch?v=LU_KD1enR3Q.

Baird, Alison. *White as the Waves: A Novel of Moby Dick*. St. John's, Nfld., Canada: Creative Book Publishing, 1999.

Balcombe, Jonathan. *What a Fish Knows: The Inner Lives of Our Underwater Cousins.* New York: Farrar, Straus and Giroux, 2016.

Bekoff, Marc. "Animal Emotions: Exploring Passionate Natures." *BioScience* 50, no. 10 (October 2000): 861–70.

Benson, John F., Peter J. Mahoney, T. Winston Vickers, Jeff A. Sikich, Paul Beier, Seth P. D. Riley, Holly B. Ernest, and Walter M. Boyce. "Extinction Vortex Dynamics of Top Predators Isolated by Urbanization." *Ecological Applications* 29, no. 3 (April 2019): e01868. https://doi.org/10.1002/eap.1868.

Benvenuti, Anne. "Evolutionary Continuity and Personhood: Legal and Therapeutic Implications of Animal Consciousness and Human Unconsciousness." *International Journal of Law and Psychiatry* 48 (September–October 2016): 43–49. https://www.sciencedirect.com/ science/article/abs/pii/S0160252716301352.

———. "Evolutionary Continuity of Personhood: Commentary on Rowlands on *Animal Personhood.*" *Animal Sentience*, no. 10 (2016): 141. https://animalstudies repository.org/cgi/viewcontent.cgi?article=1162&context=animsent.

———. *Spirit Unleashed: Reimagining Human-Animal Relations.* Eugene, Oreg.: Wipf and Stock, 2014.

Berkeley, Ellen Perry. *TNR, Past, Present, and Future: A History of the Trap-Neuter-Return Movement.* Washington, D.C.: Alley Cat Allies, 2004.

Berns, Gregory S., Andrew M. Brooks, and Mark Spivak. "Scent of the Familiar: An fMRI Study of Canine Brain Responses to Familiar and Unfamiliar Human and Dog Odors." *Behavioural Processes* 110 (January 2015): 37–46. https://doi.org/10.1016/j .beproc.2014.02.011.

Best Friends Catnippers. *Overview and Key Facts.* Flyer, n.d. http://catnippers.org/PDF Files/Overview.pdf.

Blackiston, Douglas J., Elena Silva Casey, and Martha R. Weiss. "Retention of Memory through Metamorphosis: Can a Moth Remember What It Learned As a Caterpillar?" *PLoS One* 3, no. 3 (March 2008): e1736. https://doi.org/10.1371/journal.pone .0001736.

Bonal, B. S. "Circular: Policy on Establishment of Dolphinarium." May 17, 2013. Government of India, Ministry of Environment and Forests, Central Zoo Authority. http://cza.nic.in/uploads/documents/notifications/orders/english/C-1.pdf.

Bowlby, John. *Childcare and the Growth of Love.* London: Penguin, 1953.

Boxall, Bettina. "Another Toll of the Drought: Land is Sinking Fast in San Joaquin Valley, Study Shows." *Los Angeles Times*, August 19, 2015. https://www.latimes.com /local/lanow/la-me-ln-groundwater-20150819-story.html.

Bradshaw, John. *Cat Sense: The Feline Enigma Revealed.* London: Penguin, 2014.

Brulliard, Karin. "One Problem with Shutting Down the Circus: Where Will the Animals Go?" *Washington Post*, January 19, 2017. https://www.washingtonpost.com /news/animalia/wp/2017/01/19/one-problem-with-shutting-down-the-circus -where-will-the-animals-go.

———. "What Makes Dogs So Special and Successful? Love." *Washington Post*, September 25, 2019. https://www.washingtonpost.com/science/2019/09/25/what-makes -dogs-so-special-successful-love.

Brulliard, Nicolas. "Wild Beasts of the Urban Jungle." *Park Advocate* (blog). National Parks Conservation Association, June 28, 2018. https://www.npca.org/articles /1877-wild-beasts-of-the-urban-jungle.

Bulman, May. "Villagers Knit Jumpers for Indian Elephants to Protect the Large Mammals from Near-Freezing Temperatures." *Independent*, January 19, 2017. http://www

.independent.co.uk/news/world/asia/india-elephant-jumpers-villagers-knit
-protect-near-freezing-temperatures-weather-mathura-a7535101.html.

Cacioppo, John T. "The Lethality of Loneliness." TEDx Talks. September 9, 2013. You-
Tube video. https://www.youtube.com/watch?v=_ohxlo3JoAo.

Cacioppo, John T., and William Patrick. *Loneliness: Human Nature and the Need for
Social Connection.* New York: W. W. Norton, 2008.

Call, Josep, and Michael Tomasello. "Does the Chimpanzee Have a Theory of Mind? 30
Years Later." *Trends in Cognitive Sciences* 12, no. 5 (May 2008): 187–92.

Camp Jabulani. *Camp Jabulani: An Extraordinary African Experience.* N.p.: Relais &
Châteaux, 2017. http://campjabulani.com/wp-content/uploads/2018/03/CJ
-Brochure-2017.pdf

———. "Our Elephant Herd Instrumental in Bee Research Project." September 16,
2018. https://campjabulani.com/jabulani-elephant-herd-instrumental-in-bee
-research-project.

Canon, Gabrielle. "'A Loud and Clear Message': California Passes Historic Farm Animal Protections." *Guardian*, November 8, 2018. https://www.theguardian.com
/us-news/2018/nov/08/california-animal-welfare-cage-free-eggs-prop-12-passes.

Carnie, Tony. "Why Understanding Demand for Rhino Horn Will Make or Break Their
Survival." *Business Day*, April 20, 2018. https://www.businesslive.co.za/bd
/national/science-and-environment/2018-04-20-why-understanding-demand-for
-rhino-horn-will-make-or-break-their-survival.

CBS News. "Animal Crackers' Animals 'Freed' as Boxes Get New Look." CBS News,
August 21, 2018. https://www.cbsnews.com/news/barnum-animal-crackers-box
-nabisco-mondelez-international-boxes-get-new-look-animals-freed.

Chanowitz, Benzion, and Ellen J. Langer. "Premature Cognitive Commitment." *Journal
of Personality and Social Psychology* 41, no. 6 (December 1981): 1051–63.

Christy, Bryan. "How Killing Elephants Finances Terror in Africa." *National Geographic.* August 12, 2015. https://www.nationalgeographic.com/tracking-ivory
/article.html.

Churchland, Patricia S. *Braintrust: What Neuroscience Tells Us About Morality.* Princeton, N.J.: Princeton University Press, 2011.

CNN Wire. "Formula E Car vs. Cheetah—But Which Wins Speed Race?" WHNT News.
November 28, 2017. https://whnt.com/2017/11/28/formula-e-car-vs-cheetah-but
-which-wins-speed-race.

Code, Aimee. "Neonicotinoids: Silver Bullets That Misfired." *Wings: Essays on Invertebrate Conservation* 38, no. 2 (Fall 2015): 16–21. https://xerces.org/publications
/wings-magazine/wings-382-fall-2015-rethinking-pesticides.

Colinvaux, Paul A. *Why Big Fierce Animals Are Rare: An Ecologist's Perspective.* Princeton, N.J.: Princeton University Press, 1978.

Corcoran, Kieran. "California's Economy Is Now the 5th-Biggest in the World, and
Has Overtaken the United Kingdom." *Business Insider.* May 5, 2018. https://www
.businessinsider.com/california-economy-ranks-5th-in-the-world-beating-the
-uk-2018-5?IR=T.

Cornell Lab of Ornithology. Center for Conservation Bioacoustics. https://www.birds
.cornell.edu/ccb/research.

Cornell University. "Honeybee Decision-Making Ability Rivals Any Department Committee." *ScienceDaily*, April 21, 2006. www.sciencedaily.com/releases/2006/04
/060420233908.htm.

Cunniff, Peggy, and Marcia Kramer. "Developments in Animal Law." In *The Global*

Guide to Animal Protection, edited by Andrew Linzey," 230–31. Urbana, Ill.: University of Illinois Press, 2013.

Dacke, Marie, Emily Baird, Marcus Byrne, Clarke H. Scholtz, and Eric J. Warrant. "Dung Beetles Use the Milky Way for Orientation." *Current Biology* 23, no. 4 (2013): 298–300. https://doi.org/10.1016/j.cub.2012.12.034.

Damasio, Antonio. *The Strange Order of Things: Life, Feeling, and the Making of Cultures.* New York: Pantheon, 2018.

Darwin, Charles. *The Descent of Man.* 1871. London: Penguin Classics Reprint, 2004.

"Declaration of Rights for Cetaceans: Ethical and Policy Implications of Intelligence." Symposium organized by Stephanie J. Bird and Thomas I. White, held February 19, 2012, at the annual meeting of the American Association for the Advancement of Science, Vancouver, B.C., Canada. https://aaas.confex.com/aaas/2012/web program/Session4617.html.

Dell'Amore, Christine. "Elephants Have 2000 Genes for Smell—Most Ever Found." *National Geographic Society Newsroom* (blog). National Geographic Society. July 22, 2014. https://blog.nationalgeographic.org/2014/07/22/elephants-have-2000-genes -for-smell-most-ever-found.

DiBartolomeis, Michael, Susan Kegley, Pierre Mineau, Rosemarie Radford, and Kendra Klein. "An Assessment of Acute Insecticide Toxicity Loading (AITL) of Chemical Pesticides Used on Agricultural Land in the United States." *PLoS One* 14, no. 8 (August 2019): e0220029. https://doi.org/10.1371/journal.pone.0220029.

Donaldson, Sue, and Will Kymlicka. "Farmed Animal Sanctuaries: The Heart of the Movement? A Socio-Political Perspective." *Politics and Animals* 1 (2015): 50–74.

———. *Zoopolis: A Political Theory of Animal Rights.* Oxford: Oxford University Press, 2011.

Dowling, Stephen. "The Complicated Truth about a Cat's Purr." BBC Future. July 24, 2018. https://www.bbc.com/future/article/20180724-the-complicated-truth-about -a-cats-purr.

Doyle, Brian. *Martin Marten.* New York: Picador, 2016.

———. *The Plover.* New York: Picador, 2015.

Earthfire Institute. "Dr. Con Slobodchikoff: Science and Empathy." Earthfire Institute newsletter (Winter 2015), 8–9.

Editorial Multimedia, prod. "Laguna San Ignacio—Don Pachico, the 'Grandfather of Whales.'" October 23, 2013. YouTube video. https://www.youtube.com/watch?v= R4eJTtAkD9A.

el Jundi, Basil, Emily Baird, Marcus J. Byrne, and Marie Dacke. "The Brain behind Straight-Line Orientation in Dung Beetles." *Journal of Experimental Biology* 222 (2019): 1–7. https://doi.org/10.1242/jeb.192450.

English, Colleen. "Telling the Bees." *JSTOR Daily*, September 5, 2018. https://daily.jstor .org/telling-the-bees.

Farm Sanctuary. "Julia: An Incredible Pig." March 16, 2019. YouTube video. https:// www.youtube.com/watch?v=E5lu1FRZG0.

Fenske, Michaela. "Narrating the Swarm: Changing Metanarratives in Times of Crisis." *Narrative Culture* 4, no. 2 (Fall 2017): 130–52.

Fields, R. Douglas. "Are Whales Smarter Than We Are?" *Mind Matters* (blog). *Scientific American*, January 15, 2008. https://blogs.scientificamerican.com/news-blog/are -whales-smarter-than-we-are.

Fitch, W. Tecumseh. "The Evolution of Language: A Comparative Review." *Biology and Philosophy* 20, no. 2–3 (March 2005): 193–230.

Fowler, Karen Joy. *We Are All Completely Beside Ourselves.* New York: Putnam, 2013.

Frost, Robert. *Collected Poems, Prose, and Plays.* New York: Library of America, 1995.

Gilford, Gwynn. "Why Does a Rhino Horn Cost $300,000? Because Vietnam Thinks It Cures Cancer and Hangovers." *Atlantic*, May 2013. https://www.theatlantic.com /business/archive/2013/05/why-does-a-rhino-horn-cost-300-000-because -vietnam-thinks-it-cures-cancer-and-hangovers/275881.

Gilmour, Jared. "'Fleas Were Eating Animals Alive:' California Zoo Lets Animals Slowly Die says PETA." *Sacramento* Bee, July 10, 2019. https://www.sacbee.com/news /california/article232511652.html.

Goodall, Jane, and Marc Bekoff. *The Ten Trusts: What We Must Do to Care For the Animals We Love.* New York: HarperOne, 2003.

Gooden, Lezla. "The California Living Museum Accused by PETA for Animal Cruelty, KCSO Investigating." 23ABC News [Bakersfield, Calif.]. July 11, 2019. https://www .turnto23.com/news/local-news/the-california-living-museum-accused-by-peta -for-animal-cruelty-kcso-investigating.

Gowdy, Barbara. *The White Bone.* New York: Picador, 2000.

Griffin, Donald. *The Question of Animal Awareness: Evolutionary Continuity of Mental Experience.* Los Altos, Calif.: Kaufmann, 1976.

Hallmann, Caspar A., Martin Sorg, Eelke Jongejans, Henk Siepel, Nick Hofland, Heinz Schwan, Werner Stenmans et al. "More than 75 Percent Decline over 27 Years in Total Flying Insect Biomass in Protected Areas." *PLoS One* 12, no. 10 (October 2017): e0185809. https://doi.org/10.1371/journal.pone.0185809.

Harding, Kelli. *The Rabbit Effect: Live Longer, Happier, and Healthier with the Groundbreaking Science of Kindness.* New York: Atria, 2019.

Harlow, H. F. "The Nature of Love." *American Psychology* 13, no. 12 (1958): 673–85.

Heinrich, Bernd. *Life Everlasting: The Animal Way of Death.* Boston: Mariner, 2013.

Hobgood-Oster, Laura. *A Dog's History of the World: Canines and the Domestication of Humans.* Waco, Tex.: Baylor University Press, 2014.

Hoedspruit Endangered Species Centre. "The Cheetah Breeding Project." February 7, 2018. https://hesc.co.za/2018/02/the-cheetah-breeding-project.

———. "HESC Founder Lente Roode Awarded for Her Contribution to Conservation." October 5, 2016. http://hesc.co.za/2016/10/hesc-founder-lente-roode-awarded-for -her-contribution-to-conservation.

———. "Ike, a Strong Rhino Bull, Another Victim to Victor, at HESC." September 19, 2018. https://hesc.co.za/2018/09/ike-the-strong-rhino-bull-who-survived-a -poaching-attack.

Hof, Patrick R., and Estel Van Der Gucht. "Structure of the Cerebral Cortex of the Humpback Whale, *Megaptera novaeangliae* (Cetacea, Mysticeti, Balaenopteridae)." *Anatomical Record* 290, no. 1 (January 2007): 1–31. https://doi.org/10.1002/ar.20407.

Holdgate, Martin. *John Herbert Strutt, 1935–2010: A Memoir.* Kirkby Stephen, England: Cerberus, 2011.

Hölldobler, Bert, and Edward O. Wilson. *The Ants.* Boston: Belknap, 1990.

———. *The Superorganism: The Beauty, Elegance, and Strangeness of Insect Societies.* New York: W. W. Norton, 2008.

Hopwood, Jennifer, Aimee Code, Mace Vaughan, and Scott Hoffman Black. *Recommendations to Protect Pollinators from Neonicotinoids: Suggestions of Policy Solutions, Risk Assessment, Research, and Mitigation.* Portland, Oreg.: Xerces Society for Invertebrate Conservation, 2016.

Johnson, Elizabeth A. *Ask the Beasts: Darwin and the God of Love*. London: Blooms-
   bury, 2014.
Kemmerer, Lisa. "Nooz: Ending Zoo Exploitation." In *Metamorphoses of the Zoo: Ani-
   mal Encounter after Noah*, edited by Ralph R. Acampora, 37–56. New York: Lexing-
   ton, 2010.
Kolbert, Elizabeth. *The Sixth Extinction: An Unnatural History*. New York: Picador, 2014.
Kingsolver, Barbara. *Prodigal Summer*. New York: Harper Perennial, 2001.
Kotrschal, Kurt, Jon Day, Sandra McCune, and Manuela Wedl. "Human and Cat Per-
   sonalities: Building the Bond from Both Sides." In *The Domestic Cat: The Biology of
   Its Behaviour*, edited by Dennis C. Turner and Patrick Bateson, 113–27. Cambridge:
   Cambridge University Press, 2014.
La Ganga, Maria L. "Prisoners Rescue Deaf Dogs Evacuating California Fire: 'We
   Sprung into Action.'" *Guardian*, August 3, 2016. https://www.theguardian.com
   /lifeandstyle/2016/aug/03/deaf-dogs-sand-wildfire-california-prison-shelter.
Leahy, Stephen. "Insect 'Apocalypse' in the U.S. Driven by 50x Increase in Toxic Pesti-
   cides." *National Geographic*. August 6, 2019. https://www.nationalgeographic.com
   /environment/2019/08/insect-apocalypse-under-way-toxic-pesticides-agriculture.
Levine, Peter A. *Waking the Tiger, Healing Trauma: The Innate Capacity to Transform
   Overwhelming Experiences*. Berkeley, Calif.: North Atlantic, 1997.
Linz, G. M., C. E. Knittle, and R. E. Johnson. *Ecology of Corvids in the Vicinity of the Aliso
   Creek California Least Tern Colony, Camp Pendleton, California*. Fargo, N.D.: USDA,
   North Dakota Field Station, 1990.
Loss, Scott R., Tom Will, and Peter P. Marra. "The Impact of Free-Ranging Domestic
   Cats on Wildlife of the United States." *Nature Communications* 4, 1396 (2013): 1396.
   https://doi.org/10.1038/ncomms2380.
Low, Philip. "The Cambridge Declaration on Consciousness." Edited by Jaak Panksepp,
   Diana Reiss, David Edelman, Bruno VanSwinderen, Philip Low, and Christof Koch.
   Presented and adopted July 7, 2012, at the Francis Crick Memorial Conference on
   Consciousness in Human and Non-Human Animals, Churchill College, University
   of Cambridge. fcmconference.org/img/CambridgeDeclarationOnConsciousness
   .pdf.
Lundgren, Jonathan G., and Scott W. Fausti. "Trading Biodiversity for Pest Problems."
   Science Advances 1, no. 6 (July 2015): e1500558. https://www.ncbi.nlm.nih.gov
   /pubmed/26601223.
Lyons, Leslie A. "Why Do Cats Purr?" *Scientific American*, April 3, 2006. https://www
   .scientificamerican.com/article/why-do-cats-purr.
Main, Douglas. "Why Koko the Gorilla Mattered." *National Geographic*. June 21, 2018.
   https://www.nationalgeographic.com/news/2018/06/gorillas-koko-sign-language
   -culture-animals.
Marino, Lori, Richard C. Connor, R. Ewan Fordyce, Louis M. Herman, Patrick R. Hof,
   Louis Lefebvre, David Lusseau, et al. "Cetaceans Have Complex Brains for Com-
   plex Cognition." *PLoS Biology* 5, no. 5 (May 2007): 966–72.
Marzuff, John M., and Tony Angell. *In the Company of Crows and Ravens*. New Haven:
   Yale University Press, 2005.
McGraw, Carol. "Gorilla's Pets: Koko Mourns Kitten's Death." *Los Angeles Times*, Janu-
   ary 10, 1985. https://www.latimes.com/archives/la-xpm-1985-01-10-mn-9038-story
   .html.
McNeil, Donald G., Jr. "An Invasive New Tick is Spreading in the U.S." *New York Times*,

August 6, 2018. https://www.nytimes.com/2018/08/06/health/asian-long-horned
-tick.html.

McQuay, Bill, and Christopher Joyce. "It Took a Musician's Ear to Decode the Complex
Song in Whale Calls." National Public Radio. August 6, 2015. https://www.npr.org
/2015/08/06/427851306/it-took-a-musicians-ear-to-decode-the-complex-song-in
-whale-calls?t=1571504100656.

Mercy for Animals. *Farm to Fridge* (2011). https://www.imdb.com/title/tt1863233.

Mill, John Stuart. "Coleridge." 1840. In John Stuart Mill and Jeremy Bentham, *Utilitari-
anism and Other Essays*, edited by Alan Ryan. London: Penguin Classics, 1987.

Mogomotsi, Goemeone E. J., and Patricia Kefilwe Madigele. "Live By the Gun, Die By
the Gun: Botswana's 'Shoot to Kill' Policy as an Anti-Poaching Strategy." *South African Crime Quarterly*, no. 60 (June 2017): 51–59. http://www.scielo.org.za/scielo.php
?script=sci_arttext&pid=S1991-38772017000200007.

Monkey World Ape Rescue Center. Orientation brochure (undated).

Morell, Virginia. *Animal Wise: The Thoughts and Emotions of Our Fellow Creatures*. New
York: Crown, 2013.

Mychalejko, Cyril. "Ecuador's Constitution Gives Rights to Nature." Upside Down
World. September 25, 2008. http://upsidedownworld.org/archives/ecuador
/ecuadors-constitution-gives-rights-to-nature.

National Geographic Society. "Black Mamba." National Geographic Society website,
"Animals" section. N.d. https://www.nationalgeographic.com/animals/reptiles/b
/black-mamba.

Nightingale, Florence. *Notes on Nursing: What It Is, and What It Is Not*. London: Harri-
son, 1859.

O'Connell, Caitlin. *The Elephant's Secret Sense: The Hidden Life of the Wild Herds of
Africa*. Chicago: University of Chicago Press, 2008.

———. "The Secret Lives of Elephants." *Lecture given May 13, 2014, at the* Jewish Community Center of San Francisco. September 14, 2014. YouTube video. https://www
.youtube.com/watch?v=4rdLwVBB7cI.

Oerlemans, Natasja, ed. *Living Planet Report 2016: Risk and Resilience in a New Era*.
Gland, Switzerland: World Wide Fund for Nature, 2016.

Oliver, Mary. "Wild Geese." In *Dream Work*. New York: Atlantic Monthly, 1986.

Olkowicz, Seweryn, Martin Kocourek, Radek K. Lučan, Michal Porteš, W. Tecumseh
Fitch, Suzana Herculano-Houzel, and Pavel Němec. "Birds Have Primate-Like
Numbers of Neurons in the Forebrain." *Proceedings of the National Academy of Sci-
ences* 113, no. 26 (June 2016): 7255–60.

Orr, Deborah. "The Tale of a Constipated Goldfish Has Raised My Hopes for 2105."
Guardian, January 2, 2015. https://www.theguardian.com/commentisfree/2015/jan
/02/tale-of-a-constipated-goldfish-raised-my-hopes-for-2015.

Overman, Michelle. "What Is Disorganized Attachment?" eCounseling.com. October
21, 2018. https://www.e-counseling.com/relationships/what-is-disorganized
-attachment.

Panksepp, Jaak. "Rats Laugh When You Tickle Them." Free Science Lectures. June 12,
2007. YouTube video. https://www.youtube.com/watch?v=j-admRGFVNM.

Panksepp, Jaak, and Lucy Biven. *The Archaeology of Mind: Neuroevolutionary Origins of
Human Emotions*. New York: W. W. Norton, 2012.

Paull, Laline. *The Bees*. New York: Ecco, 2015.

Payne, Roger. *Among Whales*. New York: Delta, 1995.

Pepperberg, Irene M. *Alex and Me: How A Scientist and a Parrot Discovered a Hidden World of Animal Intelligence—and Formed a Deep Bond in the Process*. New York: HarperCollins, 2008.

Pert, Candace B. *Molecules of Emotion: The Science behind Mind-Body Medicine*. New York: Simon and Schuster, 1999.

Peterson, Anna. "Religious Studies and the Animal Turn." *History of Religions* 56, no. 2 (November 2016): 232–45.

Pierce, Jessica. "Why Do Cats Meow at Humans? Understanding Your Cat's Communications." *All Dogs Go to Heaven* (blog). *Psychology Today*, September 5, 2018. https://www.psychologytoday.com/intl/blog/all-dogs-go-heaven/201809/why-do-cats-meow-humans.

Potter, Alice, and Daniel Simon Mills. "Domestic Cats (*Felis silvestris catus*) Do Not Show Signs of Secure Attachment to Their Owners." *PLoS One* 10, no. 9 (September 2015): e0135109. https://doi.org/10.1371/journal.pone.0135109.

Public Broadcasting Service. *Koko: The Gorilla Who Talks*. PBS documentary (2016). https://www.pbs.org/show/koko-gorilla-who-talks.

Ram, Ashish. "The Power of One." PoemHunter.com. January 5, 2007; last edited December 18, 2010. https://www.poemhunter.com/poem/the-power-of-one.

Redmond, Layne, and Tadeu Mascarenhas. *Hymns from the Hive*. Golden Seed Music, 2011.

Rich, Adrienne. *A Wild Patience Has Taken Me This Far: Poems 1978–1981*. New York: W. W. Norton, 1993.

Riddle, Sharla. "How Bees See and Why It Matters." *Bee Culture*. May 20, 2016. https://www.beeculture.com/bees-see-matters.

Rizzolatti, Giacomo, Maddalena Fabbri-Destro, and Luigi Cattaneo. "Mirror Neurons and Their Clinical Relevance." *Nature Clinical Practice Neurology* 5, no. 1 (January 2009): 24–34. https://doi.org/10.1038/ncpneuro0990.

Rizzolatti, Giacomo, and Corrado Sinigaglia. *Mirrors in the Brain: How Our Minds Share Actions, Emotions, and Experience*. Oxford: Oxford University Press, 2008.

Royal Society for the Protection of Birds. "Are Cats Causing Bird Declines?" N.d. https://www.rspb.org.uk/birds-and-wildlife/advice/gardening-for-wildlife/animal-deterrents/cats-and-garden-birds/are-cats-causing-bird-declines.

Rowlands, Mark. "Are Animals Persons?" *Animal Sentience*, no. 10 (2016): 101. https://animalstudiesrepository.org/animsent/vol1/iss10/1.

———. *Can Animals Be Persons?* New York: Oxford University Press, 2019.

Rumi [Jalal ad-Din Muhammad]. *The Essential Rumi*. Translated by Coleman Barks. New York: HarperOne, 2004.

Saint-Exupery, Antoine de. *The Little Prince*. New York: Mariner, 2000.

Sanchez, Olivia. "California to Build World's Largest Highway Overpass for Wildlife—Not Cars—above U.S. 101." *USA Today*, August 23, 2019. https://eu.usatoday.com/story/news/nation/2019/08/23/wildlife-crossing-ensure-future-mountain-lions/2068349001.

Saucier-Bouffard, Carl. "The Legal Rights of Great Apes." In *The Global Guide to Animal Protection*, edited by Andrew Linzey, 235–37. Urbana, Ill.: University of Illinois Press, 2013.

Scammon, Charles M. *The Marine Mammals of the North-western Coast of North America, Described and Illustrated; Together with an Account of the American Whale-Fishery*. San Francisco: John H. Carmany, 1874. https://doi.org/10.5962/bhl.title.16244.

Schneider, Mike, and Jennifer Kay. "Tilikum, Orca That Killed Trainer, Dies at Sea-World Orlando." Associated Press, January 6, 2017. https://apnews.com/doaaf30774a847d3a2d6454c3dd4598f.

Schore, Allan N. "Effects of a Secure Attachment Relationship on Right Brain Development, Affect Regulation, and Infant Mental Health." *Infant Mental Health Journal* 22, no. 1–2 (January/April 2001): 7–66.

Schwarz, Dorothy. "Free-Range Parrots: A Meeting with John Strutt in Cumbria." Cityparrots.org. January 7, 2007. Originally published in *Parrots* (UK), October 2001. http://cityparrots.org/journal/2007/1/7/free-range-parrots-a-meeting-with-john-strutt-in-cumbria.html.

Seeley, Thomas D., Scott Camazine, and James Sneyd. "Collective Decision-Making in Honey Bees: How Colonies Choose among Nectar Sources." *Behavioral Ecology and Sociobiology* 28, no. 4 (April 1991): 277–90.

Senese, Vincenzo Paolo, Simona De Falco, Marc H. Bornstein, Andrea Caria, Simona Buffolino, and Paola Venuti. "Human Infant Faces Provoke Implicit Positive Affective Responses in Parents and Non-parents Alike." *PLoS One* 8, no. 11 (November 2013): e80379. https://doi.org/10.1371/journal.pone.0080379.

Sewall, Katy. "The Girl Who Gets Gifts from Birds." *BBC Magazine*. February 25, 2015. https://www.bbc.com/news/magazine-31604026.

Shamoo, Adil E., and David B. Resnik. *Responsible Conduct of Research*. New York: Oxford University Press, 2009.

Shubin, Neil. *Your Inner Fish: A Journey into the 3.5-Billion-Year History of the Human Body*. New York: Vintage, 2009.

Siebert, Charles. "Watching Whales Watching Us." *New York Times*, July 8, 2009. https://www.nytimes.com/2009/07/12/magazine/12whales-t.html.

Siegel, Alan. *The Neurobiology of Aggression and Rage*. Boca Raton, Fla.: CRC, 2004.

Singer, Peter. *Animal Liberation: A New Ethics for Our Treatment of Animals*. London: Jonathan Cape, 1976.

Slobodchikoff, Con. *Chasing Doctor Dolittle: Learning the Language of Animals*. New York: St. Martin's, 2012.

Smithsonian National Zoological Park. "Shanthi, the National Zoo's Musical Elephant, Plays the Harmonica." May 1, 2012. YouTube video. https://www.youtube.com/watch?v=FgCu84NPbnw.

Solly, Meilan. "California Will Build the Largest Wildlife Crossing in the World." *Smithsonian*, August 21, 2019. https://www.smithsonianmag.com/smart-news/california-will-build-largest-wildlife-crossing-world-180972947.

Spitz, R. A., and K. M. Wolf. "Anaclitic Depression: an Inquiry into the Genesis of Psychiatric Conditions in Early Childhood." *Psychoanalytic Study of the Child* 2 (1946): 313–42.

Spottiswoode, Claire N., Keith S. Begg, and Colleen M. Begg. "Reciprocal Signaling in Honeyguide-Human Mutualism." *Science* 353, no. 6297 (July 2016): 387–89. https://www.ncbi.nlm.nih.gov/pubmed/27463674.

Sullivan, Katherine. "Historic Victory! 3 PETA-Backed Animal-Friendly California Bills Signed into Law." PETA blog, last updated October 22, 2019. https://www.peta.org/blog/new-california-laws-victories-animals.

Tautz, Jürgen. *The Buzz about Bees: Biology of a Superorganism*. Berlin: Springer-Verlag, 2008.

Thomas, Elizabeth Marshall. *The Tribe of Tiger: Cats and Their Culture*. New York: Simon and Schuster, 1994.

Tirado, Reyes, Gergely Simon, and Paul Johnston. *Bees in Decline: A Review of Factors That Put Pollinators and Agriculture in Europe at Risk*. Greenpeace Research Laboratories Technical Report (Review) 01-2013. Amsterdam, Netherlands: Greenpeace International, 2013. https://www.greenpeace.to/greenpeace/?p=1543.

Tucker, Marlee A., Katrin Böhning-Gaese, William F. Fagan, John M. Fryxell, Bram Van Moorter, Susan C. Alberts, Abdullahi H. Ali et al. "Moving in the Anthropocene: Global Reductions in Terrestrial Mammal Movements." *Science* 359, no. 6374 (January 2018): 466–69. https://doi.org/10.1126/science.aam9712.

Vitale, Kristyn R., Alexandra C. Behnke, and Monique A. R. Udell. "Attachment Bonds between Domestic Cats and Humans." *Current Biology* 29, no. 18 (September 23, 2019): 864–65.

Walker, Alice. *The Chicken Chronicles: Sitting with the Angels Who Have Returned with My Memories; Glorious, Rufus, Gertrude Stein, Splendor, Hortensia, Agnes of God, the Gladyses, & Babe; A Memoir*. New York: New Press, 2011.

Watson, John Broadus, and Rosalie Rayner Watson. *Psychological Care of Infant and Child*. New York: W. W. Norton, 1928.

Wedl, Manuela, Barbara Bauer, Dorothy Gracey, Christine Grabmayer, Elisabeth Spielauer, Jon Day, and Kurt Kotrschal. "Factors Influencing the Temporal Patterns of Dyadic Behaviours and Interactions between Domestic Cats and Their Owners." *Behavioural Processes* 86 (2011): 58–67.

Weiss, Alexander, James E. King, Miho Inoue-Murayama, Tetsuro Matsuzawa, and Andrew J. Oswald. "Evidence for a Midlife Crisis in Great Apes Consistent with the U-Shape in Human Well-Being." *Proceedings of the National Academy of Sciences* 109, no. 49 (December 2012): 19949–52. https://doi.org/10.1073/pnas.1212592109.

Wilson, Edward O. *Half-Earth: Our Planet's Fight for Life*. New York: Liveright, 2017.

Winnicott, Donald. *The Child, the Family, and the Outside World*. London: Penguin, 1947.

Winter, L. "Court Declares Captive Orangutan Is 'Non-Human Person.'" IFLScience, n.d. (2014). http://www.iflscience.com/plants-and-animals/argentinian-court-extends-rights-captive-orangutan.

Wynne, Clive D. L. *Dog is Love: Why and How Your Dog Loves You*. New York: Houghton Mifflin Harcourt, 2019.

Zimmer, Carl. "To Bee." *National Geographic*, October 25, 2006. https://www.nationalgeographic.com/science/phenomena/2006/10/25/to-bee.

INDEX

Adams, Sharon, 166
ADHD (attention-deficit/hyperactivity disorder), 65
affective neuroscience, 34–36, 86, 153–54. *See also* Panksepp, Jaak
affective systems, 35, 55, 131, 153–54, 190, 212n7; care, 35, 55, 57–58, 109, 136; fear, 35, 86–88; grief, 34–37; lust, 35, 54–58; play, 35, 64–66; rage, 35, 87–88, 139–41, 143, 153; seeking, 35, 55, 79–80, 114–15, 165, 208
agouti, 17
Akhtar, Aysha, 28
Alaska Humane Society, 149
Alex (parrot), 156
All-Ball (cat), 173
Alley Cat Allies, 129
alligator, 68, 183
Alpha (cat), 111–15, 119, 120
Amazon River, 17, 30
American Academy of Sciences, 182
amphibian, 13, 129, 130, 198
anatomy: of bee, 190; of black mamba, 47; of butterfly, 188; cell, 188; of cheetah, 79–80; of elephant, 62–63, 65; imaginal disc, 188, 189, 210; of insect, 190; of raven, 3–5, 7; of rhino, 51, 85; shared fundamentals of, 3, 5, 79–80, 82, 83, 131; of whale, 80, 92, 93, 102–3, 107. *See also* brain

Aneen (ranger), 11, 18, 52–54, 77; birds and, 52–53
animal abuse, 13–14, 29, 160, 178, 181; in tourism, 59, 68–69
Animal Adventure Park, 57, 212n9
animal ambassador, 6, 24, 30, 131, 162–63, 168; Koko (gorilla), 172–74
Animal Behavior Society, 65
animal cognition, 81, 93, 94
animal communication, 80–81, 142, 180, 186; within bee colony, 191; of cat to human, 112–13, 116–17; elephant seismic, 11, 53, 63, 81; muscle mimicry, 142, 213n13 (chap. 5); via urine or feces, 82, 85, 114, 116; of whale, 181. *See also* language
animal consciousness, 179–80
animal conservation, 45, 134, 155, 167; Center for Elephant Conservation, 180; Centre for Parrot Conservation, 155–56; at Kapama, 42, 44, 50, 77; line between rescue and, 11; programs, 42, 50, 175–76; success stories, 96; Wildlife SOS Elephant Conservation and Care Centre, 178; at zoos, 165. *See also* HESC; Kapama Private Game Reserve
animal rescue, 14–15, 24, 36, 134, 176, 190; dog rescue, 138–39, 143–47; Furever Dachshund Rescue, 146; at Kapama, 50,

animal rescue (*continued*)
    60–61; line between conservation and,
    11; Monkey World Ape Rescue Centre,
    159–61, 185; no-kill shelter, 129, 147–49;
    program, 59, 70, 72–73, 77, 129, 175–76;
    Three Rivers Humane Society, 148–50;
    wildlife, 6, 7, 166. *See also* CALM; Camp
    Jabulani; HESC
animal rights, 23–24, 74, 182, 183, 186–87
animal sanctuary, 73, 175–76; farm
    sanctuary, 23, 24; Farm Sanctuary (N.Y.),
    20–39 passim, 176; Global Federation
    of Animal Sanctuaries, 180; ocean
    sanctuary, 96, 180, 182, 186; wildlife
    sanctuary, 6, 169. *See also* animal
    conservation; animal rescue
animal welfare. *See under* legal system
Animal Welfare Institute, 180
Anneliese (Anne), 98–109 passim
ant, 168, 177, 194, 195, 196–97; carpenter, 198
antelope, 9
anthropomorphism, 65–66
ape, 15, 17, 161, 173–74, 182; Monkey World
    Ape Rescue Centre, 159–61, 185
April (giraffe), 57
Argentina, 182
Aridjis, Homero, 102
art, animal in, 183–84
attachment, 190; of cat, 118, 119, 144; of dog,
    119, 135–37, 144; neurobiology of, 190; style,
    118, 119, 144
attention-deficit/hyperactivity disorder
    (ADHD), 65

baboon, 45, 67, 68, 181
badger, 171
Baird, Allison, 184
Baja California Sur, 92, 95, 103; birthing
    grounds, 17, 93, 95, 176; lagoons, 93–94,
    95–96, 97, 99–100, 106
Bancroft, Anne, 138
Barbilla (whale), 99, 107
bear, 67, 68, 166, 170, 171, 187
beaver, 56
bee, 72, 82, 103, 142, 184; bumblebee, 199;
    mason, 200; worship of, 206. *See also*
    honeybee; human-animal relations
beetle: dung, 40–42, 50, 198; saw-toothed
    grain, 198

Bekoff, Marc, 17, 37, 65
Benedict Goat, 20, 32–33, 39
Best Friends Catnippers, 121, 122, 129
Betty (animal shelter friend), 148–51
Bhutan, 186
Bible, 16, 33
big five, 40–55 passim, 168
biological science, 16, 80, 178–79
biophilia, 161, 193, 208. *See also* human-
    animal relations
bird, 2–3, 15, 74, 129–30, 158–59; in
    Africa, 18–19, 52–53, 206; brain of, 4, 5;
    budgerigar, 158; buzzard, 9; corn bunting,
    130; goose, 171; Hawaiian cardinal, 6;
    honeyguide, 206; hornbill, 53; magpie,
    209; migratory, 80; oxpecker, 85; plate-
    billed mountain toucan, 187; rescue, 149,
    166; Royal Society for the Protection of
    Birds, 130; seabird, 102, 104, 108; sleep of,
    5; sparrow, 1, 74; swan, 56; tree sparrow,
    130; vulture, 90; woodpecker, 6. *See also*
    chicken; parrot; raven; songbird; turkey
    —specific birds: Alex (parrot), 156;
    Emma (Hawaiian cardinal), 5–6; Fling
    (woodpecker), 6. *See also* Jason; Peanut
bison, 170, 171. *See also* buffalo
Biven, Lucy, 35, 57, 65–66, 85–87, 114, 140, 165.
    See also affective systems; Panksepp, Jaak
black mamba, 46–47, 50–51
bobcat, 67, 166, 171
Botswana, 89
Bradshaw, John, 117, 119
brain, 35, 88; amygdala, 86–87, 141; of
    bird, 4, 5; of cat, 141; dopamine, 114; of
    dung beetle, 41; endogenous opioid,
    141; endorphin, 114, 192; evolution of,
    35; forebrain, 4; gyrification index,
    94; hypothalamus, 141; of insect,
    190; mammalian, 35, 55, 64, 114, 139–
    40, 165; mirror neuron, 142, 213n13
    (chap. 5); neurochemical, 35, 55;
    neuromodulator, 141; neuron, 4, 191, 197,
    209; neurotransmitter, 64, 65, 132, 139, 192;
    periaqueductal gray, 141; of primate, 4,
    209; of raven, 4, 5; reward center, 135; of
    rhino, 85; similarity or difference across
    species, 35, 142, 190; spindle neuron, 93;
    substrate, 179–80; of whale, 93, 94–95, 110.
    See also affective neuroscience; affective

systems; oxytocin; social neuroscience;
     substance P
Bronwyn (dog), 137–40, 143–44, 153
Brown, Tom, 122
buffalo, 9, 10, 11, 44; big five, as one of, 40;
     cape, 40; water, 45
bug, 12, 58, 191, 195, 196–98, 210; sow, 194;
     superbug, 16, 195. *See also* insect
bush baby, 46
butterfly, 187, 188, 195; monarch, 198

Cacioppo, John, 36, 134–35, 191–92, 217n5
Caforio, Roberto, 202, 204–5, 207, 208
caiman, 183
California Living Museum Zoo. *See* CALM
California State University, Northridge. *See*
     CSUN
Callie (cat), 120
CALM (California Living Museum Zoo), 6–7,
     8, 165–67
Camp Jabulani, 43–44, 59, 61–62, 68–77
     passim
Cape turtledove, 67
care. *See* affective systems
cat, 17, 27, 111–33 passim, 135, 149, 176; Alley
     Cat Allies, 129; attachment style, 118, 119,
     144; big, 132, 164–65, 167; bird and, 112, 121,
     129–30; brain of, 141; domestic (house), 111,
     112, 118–19; feral, 2, 15, 112–15, 119, 120–32;
     fisher, 166; human relations and, 144, 116–
     19, 209; hunting by, 112, 119, 121, 124, 129–
     30; kitten, 111–12, 120, 126; lynx, 171; mouse
     and, 115–16, 124, 128, 133; puma, 132; purr
     of, 113, 116–17, 123–24, 131; rage experiment,
     141; Stray Cat Alliance, 129; TNR (trap-
     neuter-return), 119–32 passim. *See also*
     animal communication; bobcat; cheetah;
     CSUN: CSUN Cat People; death; Hyde
     Park Cats; leopard; lion
—specific cats: All-Ball, 173; Callie, 120;
     Figaro, 123–24; Idgie, 131–32; JouJou, 111,
     112; Mac, 120; Olive, 131–32; Stella, 111–12.
     See also Alpha; Centaura
caterpillar, 187–88
Catnippers, Best Friends, 121, 122, 129
Cecil (lion), 14
Centaura (cat), 111–15, 119, 120, 132–33
Center for Elephant Conservation, 180
Centre for Parrot Conservation, 155–56

chameleon, 48, 58
"Charlie bit me," 103, 159, 196
Chauvet Cave, 136
cheetah, 77–80, 83, 90, 130, 136; anatomy of,
     79; breeding of, 77–78, 176; cub, 44, 77–78,
     79; genetic diversity, 77–78; Sebeka, 44, 58.
     See also endangered species; HESC
chicken, 22, 26, 30–31, 39, 109, 171; broiler,
     27–28
chimpanzee, 160, 181, 184, 185; kidnapped,
     161; midlife crisis in, 165
Chukchi Sea, 95, 96
Churchland, Patricia, 142
circus, 161, 162–63, 176, 181; animal release
     from, 180; ethics of, 162–63, 164–65; legal
     regulation of, 183
CITES (Convention on International Trade
     in Endangered Species of Wild Fauna and
     Flora), 59, 79, 83
civet, 56, 58
climate change, 10–13, 45, 177. *See also*
     drought; ecological destruction
Cloud Atlas (film), 31
cockroach, 142
Coetzee, Paul, 59–63, 67, 69
coevolution, 7, 136, 192, 206
Cohn, Ron, 172
Colinvaux, Paul, 58
companion animals. *See under* human-
     animal relations
conservation. *See* animal conservation
Convention on International Trade in
     Endangered Species of Wild Fauna and
     Flora. *See* CITES
coronavirus, 16
Coston, Susie, 22, 25–33, 38, 39
cougar, 171, 172. *See also* mountain lion
Coulombe, Laura, 144–47
cow, 22, 28, 30, 31, 36–37, 109; Angus, 34–35;
     Frank Cow, 33–34, 39; Holstein, 27; *The
     Innocent Eye Test*, 183
coyote, 3, 7, 67, 123, 171, 184
crab, 167
crocodile, 54, 183
crustacean, 190
CSUN (California State University,
     Northridge), 119–23, 125, 127–30; CSUN
     Cat People, 120, 122, 125, 127–29
Cunniff, Peggy, 182

Damasio, Antonio, 36
Darwin, Charles, 81, 88, 94
death, 7, 29–30, 39, 41, 102, 188; of bees, 103,
    190; of birds, 7, 130; by disease, 150; in
    drought, 11, 45; of feral cats, 115, 122, 123,
    128; by euthanasia, 120–21, 129, 146; by
    flea infestation, 166; from grief, 37, 141;
    in laboratory, 193; by slaughter, 18, 21, 28,
    30–31, 33–34, 178; by starvation, 45, 166; of
    young, 26, 77, 79, 188. *See also* extinction;
    hunting; poaching
Debevere, Marjan, 116
Declaration of Rights for Cetaceans, 182
deer, 171. *See also* hunting
depletion of wildlife, 8–9, 10–11, 12, 71, 89–90.
    See also under prey
Descartes, 56, 81, 192
differences of degree, 81–83
dog, 14–15, 17, 27, 134–54 passim, 171, 176;
    African wild, 77, 90; attachment style,
    119, 135–37, 144; Australian shepherd, 148;
    beagle, 151–52; bulldog, 27; Chihuahua,
    27; dachshund, 145–47; evolution of, 136;
    feral, 113, 114; Furever Dachshund Rescue,
    146; genetic diversity of, 136; Italian
    greyhound, 145; jackal, 46; Jack Russell
    terrier, 2, 137–38, 143; puppy, 112, 134, 138–
    39, 145, 150, 151; springer spaniel, 134;
    Three Rivers Humane Society, 148–50;
    training, 143. *See also* animal rescue
—specific dogs: Eleanor Roosevelt, 134; Lexi,
    137; Molly Brown, 137–38, 143; Rusty, 134;
    Sherry, 151–54. *See also* Bronwyn
dolphin, 56, 82, 102, 180; Amazon pink river,
    17; Declaration of Rights for Cetaceans,
    182
Donaldson, Sue, 23, 26, 31
Dotty Goat, 20, 38, 39
Doyle, Brian, 184
dragonfly, 195
drought, 1, 8, 10–12, 19, 44–45, 71
Dr. Vicky, 137–40, 143
Drynan, Jerilee, 148–51
Duke, Patty, 138
dung beetle, 40–42, 50, 198

Earth, 3, 13, 55, 109, 207; beauty of, 208;
    exploitation of, 12; life on, 12, 17, 67, 81,
    154, 198; to nature, half of, 168; shared
    experience of, 86, 154, 171; from space,
    177, 187
Earthfire Institute, 169–72, 185
eastern North Pacific gray whale, 92–110
    passim, 176; anatomy of, 92, 93, 102–3;
    baby, 92, 95, 101, 104, 105, 107; Barbilla, 99,
    107; brain of, 93, 94–95, 110; Declaration of
    Rights for Cetaceans, 182; International
    Agreement for the Regulation of
    Whaling, 96; interspecies culture, 17,
    92–107 passim; mating, 96; migration of,
    92, 95, 96, 106; protection of, 93; slaughter
    of, 87, 92–93, 96–97, 99, 108. *See also* Baja
    California Sur; human-animal relations;
    International Whaling Commission;
    oxytocin; whale
ecological destruction, 16, 74, 193, 206. *See
    also* climate change; drought; habitat loss
ecological niche, 80–81, 135, 184, 201
ecosystem, 41, 42, 197, 198; bees, orientation
    to, 202–3; loss of, 77; restoration of, 167,
    168
Ecuador, 186–87
Eden Valley, 155–57, 159
Edward (ranger), 8–10, 40–51 passim, 58
Eirich, Susan, 169–71
Eleanor Roosevelt (dog), 134
elephant, 8–9, 15, 17, 42–43, 50, 59–74
    passim; acoustic cavities in feet, 11, 53, 63,
    81; African, 184; anatomy of, 62–63, 65;
    Asian, 164; baby, 61, 64, 65, 68, 70; bees
    and, 72; big five, as one of, 40; Center for
    Elephant Conservation, 180; in circus, 161,
    176–77; dung of, 41–42, 72, 75; in grief, 71;
    matriarch, 61–62; musical instruments
    and, 164; reproduction, 69–70; safari
    on back of, 43, 50, 59–61, 67–68, 69,
    70–73; in sweater, 178, 184–85; Wildlife
    SOS Elephant Conservation and Care
    Centre, 178. *See also* Camp Jabulani; ivory;
    Kapama Private Game Reserve; poaching
—specific elephants: Fishan, 60; Kumbura,
    70; Limpopo, 61, 69; Sebakwe, 69; Shanthi,
    164; Timisia, 70. *See also* Jabulani; Tokwe
El Vizcaíno Biosphere Reserve, 95
Emma (Hawaiian cardinal), 5–6
endangered species, 59, 76, 91, 198, 199;

breeding of, 77–78, 83, 136, 161, 176; U.S.
   endangered species list, 96. *See also*
   HESC
Endangered Species Act, 96
England, 23, 34, 155, 159, 189
English, Colleen, 206
entertainment, animals in, 13, 69, 162–63,
   180. *See also* circus
environmental niche, 15, 82, 85, 180
evolutionary adaptation, 15, 48, 80, 88;
   bee colony as adaptive unit, 191; human
   cooperative sociality as adaptive, 191–92;
   neoteny, 136
evolutionary biology, 15–16, 82, 83, 179
evolutionary continuity, 65, 81–82, 94, 179–
   80, 206
exotic pet, 13, 14, 46
exploitation of animals, 21, 163, 176
extinction, 12–13, 78, 96–97, 177–78, 189;
   sixth, 12, 13

Faoro, Tabita, 201–2, 203, 204, 208
farming: factory, 17, 26–30, 39; industrial,
   13, 177
farm sanctuary, 23, 24
Farm Sanctuary (N.Y.), 20–39 passim, 176
Farm to Fridge (film), 21
fear. *See* affective systems
Feld Entertainment, 176, 180
Fenske, Michaela, 194, 206, 207, 208
ferret, 149
Figaro (cat), 123–24
fish, 13, 21, 22; as food, 22, 99, 108, 113; in
   human descent, 190
Fishan (elephant), 60
Fitch, Tecumseh, 94
flea, 122, 166
Fling (woodpecker), 6
Flinkman, Debbie, 164
fly, 194, 198
food chain, 12–13, 16, 194, 197, 199. *See also*
   extinction; predator; prey
Fossie (tracker), 18–19, 51–54, 77
Fowler, Karen Joy, 184
fox, 2, 67, 115, 166, 171, 172; Loki, 170
Frank Cow, 33–34, 39
Frohoff, Toni, 106
Frost, Robert, 76

Furever Dachshund Rescue, 146
Future Farmers of America, 31

gazelle, 45
gecko, 195
genet, 46
genetic diversity: of cheetah, 77–78, 136; of
   dog, 136; of mountain lion, 167
genetic manipulation, 27–28, 39
giraffe, 9, 45, 48, 67; April, 57
Global Federation of Animal Sanctuaries,
   180
global warming, 45, 71, 96. *See also* climate
   change
gnat, 194
goat, 20, 22, 26, 32, 37–39, 74; Benedict Goat,
   20, 32–33, 39; Dotty Goat, 20, 38, 39; Skye,
   20, 32, 37, 39
goldfish, 176, 189
Gombe, 134
Goodall, Jane, 17, 37, 134
gorilla, 172–73; Gorilla Foundation, 172;
   Koko, 172–74
Gorilla Foundation, 172
Gowdy, Barbara, 184
grazing animals, 9–10, 11, 44–46, 88
grief. *See* affective systems
Guarini, Giovanni, 202–3, 204–5, 207, 208

habitat loss, 13, 78, 166–68, 177; bird
   depletion because of, 130; in coastal
   waters, 96–97; vagility, 167–68
Harding, Kelli, 169
hedgehog, 115
Heinrich, Bernd, 41
HESC (Hoedspruit Endangered Species
   Centre), 43–44, 49–50, 76–79, 83–91, 176
Hess, Tara, 25, 33
Hess, Walter, 141
hippo, 54, 58, 183
Hodgson, Richard, 158, 159
Hoedspruit Cheetah Project, 77
Hoedspruit Endangered Species Centre. *See*
   HESC
Holmes, Liezel, 9, 42–43, 52, 58; animal
   superiority and, 49, 82; anti-poaching
   and, 51, 89; climate change and, 10, 45, 71
honeybee, 190–94, 198–209 passim;

honeybee (*continued*)
  beekeeping, 199–205, 207–8; colony as
  superorganism, 191, 192, 199; DNA of, 190;
  elephants and, 72; genetic similarity with
  mammal, 190; interspecies partnership
  with, 191, 193, 194, 199–200; larvae, 199;
  population collapse, 16, 199; swarm, 193–
  94, 207; swarm as symbol, 193, 207–8, 210;
  waggle-dance, 193–94, 200–201. *See also*
  bee; human-animal relations
horse, 147, 149
Hoyle, Fred, 187
human-animal relations, 60, 66, 74–75,
  81–83, 104, 168–69; bees, mystical
  connection to, 191, 198–99, 200, 201–
  2, 203, 205–7; coevolution, 7, 136, 206;
  companion animals, 115–19, 132, 134–37,
  149; companion species, 199; courtship, 7,
  99–100, 113–14; cross-species empathy, 66,
  79–80, 82, 109, 142, 177, 186; as extended
  family, 13, 16–17, 132, 135, 177–78, 193;
  global cultural shift, 8, 14, 171, 175–78,
  187; indigenous understanding of, 17;
  interspecies love, 93, 117–18, 135, 152, 154,
  173–74; as kin, 176–77, 180–81, 185–86,
  189; shared psychology, 3, 85–86, 179–
  80; whales seeking, 92–107 passim. *See
  also* animal communication; animal
  conservation; animal rescue; animal
  sanctuary
human as distinct/superior, assumption of,
  15–16, 49, 80–81, 94, 141, 208
Humane Society, 149
hunting, 14, 44, 136, 177; of big game, 40, 186;
  of deer, 157–58; of endangered species, 77;
  of whale, 87, 92–93, 96–97, 99, 108
Hurricane Katrina, 27, 135
Hyde Park Cats, 120, 124, 127
hyena, 79

Idgie (cat), 131–32
Ike (rhino), 84–85, 86, 88, 90
India, 177, 178, 182, 184
Innocent Eye Test, The (Tansey), 183
insect, 13, 41, 43, 168, 190–99 passim, 208;
  apocalypse, 197; bees and other insects,
  204; bee as, 205, 206; diversity, loss of,
  197–98; eating of, 5, 85, 112, 119, 198; well-
  being of, 213n17 (chap. 4). *See also* bug

internal felt sense of life, 48–49, 55–56, 141–
  43, 175, 182, 197; animal consciousness,
  179–80; subjectivity, 16–17, 142, 184, 185.
  See also mind
International Agreement for the Regulation
  of Whaling, 96
International Whaling Commission, 79, 93,
  96
interspecies community, 2, 23–25, 26, 38.
  See also eastern North Pacific gray whale;
  human-animal relations
Irwin, James, 187
Italy, 47, 120, 167, 194, 209
ivory, 71, 74, 89–90, 213n24

Jabulani (elephant), 59–63, 67–69, 71, 73, 154
jackal, 46
jaguar, 187
Jain, 177
Jason (raven), 2–3, 5–8, 154, 166, 209
Jefferson County, Oreg., 149
JouJou (cat), 111, 112
Julia Pig, 28–30, 39

Kapama Private Game Reserve, 41–57
  passim, 76–77, 78, 89, 90; Buffalo Camp,
  58; Buffalo Lodge, 51; elephant rescue
  at, 59, 61, 69–70, 72; River Lodge, 40–52
  passim, 89. *See also* Camp Jabulani; HESC
Kauffman, Bill, 20–23, 25, 33, 37–39
Kauffman, Patt, 21, 22
Keller, Helen, 138
Kemmerer, Lisa, 163
Kern County, 166
kin. *See under* human-animal relations
Kingsolver, Barbara, 184
Kirkby Stephen, 155–56, 159, 176
Koditek, Diane, 6; at Kapama, 8–9, 18, 42–54
  passim, 59, 68–71, 77
Koester, Jolene, 122
Koko (gorilla), 172–74
Kramer, Marcia, 182
Kraybill, Mary Jean, 119–20, 123–27
Kruger National Park, 41, 70, 84
kudu, 9
Kumbura (elephant), 70
Kymlicka, Will, 23, 26, 31

laboratory research, 13, 163, 169, 172, 176,

193; animal release from, 180–81; ethics of, 163, 180–81; great ape banned from, 182

Laguna San Ignacio, 95, 97, 99–101, 104, 107–8. *See also* Baja California Sur

language, 15, 80–81, 94, 173–74, 185, 186; American sign, 172, 173; of bee, 193, 201, 209; of cat, 112, 113; of empathy, 142; Koko (gorilla), 172–74; linguistic capacity, 15, 80–81; of prairie dog, 172, 179; of raven, 209; Slobodchikoff and, 172, 178–79; song, 4, 61, 67, 94, 209; vocalization, 4, 181. *See also* animal communication

Lascaux, 184

legal system, 24, 130, 175, 181 83, 186–87; animal participation in, 25, 31, 183; animal welfare law, 23, 148, 182, 183; definition of personhood, 24, 176, 182–83; fur and skin sale, 183; international protection, 96–97, 160, 182; jurisdiction boundary, 78–79, 119; 148, 151; policy, 83, 89, 90, 129; Wildlife and Countryside Act, 158. *See also* animal rights; Endangered Species Act; personhood

leopard, 44–45, 167; big five, as one of, 40

Leveque, Murina, 169

Lexi (dog), 137

Limpopo (elephant), 61, 69

lion, 9–12, 42, 44, 52, 79; big five, as one of, 40; charged by, 49; mating, 54–58. *See also* mountain lion

—specific lions: Cecil, 14; P22 (mountain lion), 168

literature, animal as protagonist in, 184

Little Prince, The (Saint-Exupéry), 163

Living Planet Report, 12

lizard, 112, 133, 166, 183

Loki (silver fox), 170

London, Jack, 87

Los Angeles, 120, 123, 167, 168

Los Angeles County, 129

Louise (sheep), 20, 32

Lulu (monkey), 161

lust. *See* affective systems

lynx, 171

Mac (cat), 120

macaw. *See under* parrot

Magdalena Bay, 92, 97, 99

Magliocco, Sabina, 119–23, 125, 127–33, 136–37

Magnum (tracker), 40, 46

Main, Douglas, 172

mammal, 13, 17, 81, 168, 209; brain of, 35, 55, 64, 114, 139–40, 165; consciousness in, 180; fear in, 86, 87; genetic similarity with honeybee, 190; oxytocin in, 106, 117–18; psychology of, 85; shared fundamentals of, 35, 55, 57, 79, 83, 208; social bonds of, 36, 117, 141; young, 64, 65. *See also* affective systems

Mann, Gabi, 2

maternal bond, 36, 165, 179, 209; of cheetah, 79; of elephant, 62, 64; of whale, 95, 104, 105–7

Mathole, Isaac, 67–68

Mayoral, Francisco "Pachico," 97, 99–100, 108

Mayoral, Ranulfo, 97, 99–102, 104, 108

McCann, Lisa, 200–201, 208

meat, 14, 17–18, 61, 176; humanely raised, 37; production of, 26–31

mental life. *See* mind

Metropolitan Museum of Art, 216n26

Milky Way, 41–42

Miller, Jeffrey, 182

mind: basic features of, 80–82; hive mind, 208, 209; mind-body medicine, 192; shared features of mental life, 94, 180. *See also* affective systems; internal felt sense of life; psychology

Minnesota Zoo, 161–62

Miracle Worker, The (film), 138

Molly Brown (dog), 137–38, 143

monarch butterfly, 198

mongoose, 50–51, 53, 58

Monica (whale guide), 99, 101–2

monkey: Asian, 49, 162; brown-headed spider, 187; at Kapama, 45, 49, 51, 61, 67; Lulu, 161; at Monkey World, 160, 161; Monkey World Ape Rescue Center, 159–61, 185; vervet, 185

Monkey World Ape Rescue Centre, 159–61, 185

mosquito, 195

moth, 187

mountain lion, 67, 166, 167–68

mouse, 87, 115–16, 119, 124, 133; pocket, 168. *See also* rodent

National Aquarium, 180, 182
National Geographic, 26, 46, 71, 172, 173
National Wildlife Federation, 167
National Zoo (Smithsonian National
    Zoological Park), 164
nature, rights of, 186–87
neoteny, 136
Nerem, Robert, 169
Niemöller, Martin, 13
Nightingale, Florence, 116
Nobel Prize in Physiology or Medicine, 141
nonprofit, 83, 84, 124, 129, 144–46, 149
nyala, 45

octopus, 180
Ojo de Liebre, 95
Olive (cat), 131–32
Oliver, Mary, 215n18 (chap. 9)
orangutan, 56, 182, 185
orca, 103, 181–82
Orr, Deborah, 189
Oscine, 4. *See also* raven; songbird
oxytocin, 35–36, 57, 58, 118, 178; in dog, 135;
    in elephant, 64, 178; rage and, 141, 153;
    substance P and, 153; in whale, 106–7, 109

P22 (mountain lion), 168
Pacific gray whale. *See* eastern North Pacific
    gray whale
pangolin, 15, 43
Panksepp, Jaak, 35, 55, 64, 106; Biven and,
    35–36, 57, 65–66, 85–87, 114, 140, 165. *See
    also* affective systems
parrot, 157–59; African gray, 156, 157; Centre
    for Parrot Conservation, 155–56; macaw,
    155–56, 158, 176; Parrot Park, 155
—specific parrots: Alex, 156; Peanut, 156,
    157, 159
Patrick, William, 134–35
Patterson, Penny, 172–74
Paull, Laline, 184
Payne, Katy, 94
Payne, Roger, 94
Peanut (parrot), 156, 157, 159
People for the Ethical Treatment of Animals
    (PETA), 166, 181
Pepperberg, Irene, 156
personhood, 17, 39, 176, 182–83; of animals,
    17, 22–23, 83–86, 162–63, 172, 175–76;

of elephant, 70; nonhuman, 24, 182; of
    insect, 197; of whale, 101. *See also* legal
    system
Pert, Candace, 192–93
PETA (People for the Ethical Treatment of
    Animals), 166, 181
Peterson, Anna, 184–85
Phillipa (rhino), 83–85, 86, 88, 90, 213n11
pig, 22, 26–39 passim; Julia Pig, 28–30, 39;
    Sebastian Pig, 37, 38, 39
play. *See* affective systems
poaching, 13, 59, 77, 177; anti-, 43, 83, 89–90;
    of elephant, 71; industry, 70, 89; of rhino,
    51, 77, 83–84. *See also* Ike; Phillipa
porcupine, 171
possum, 166, 195
post-traumatic stress disorder (PTSD),
    84–89, 162
prairie dog, 172, 179
prairie vole, 56
Pratt, Beth, 168
predator, 11, 12, 44–46, 49, 58; avoidance of,
    36, 77, 79, 88; cat as, 129–30; insect as, 198
prey, 49, 58, 78, 88, 90, 112; depletion of, 45,
    129–30, 119, 197–98
primate, 45, 159–60, 185; brain of, 4, 209;
    bush baby, 46; gorilla, 172–73; Gorilla
    Foundation, 172; Monkey World Ape
    Rescue Centre, 159–61, 185; smuggling
    of, 160, 161; in zoo, 162, 163, 165. *See also*
    ape; baboon; chimpanzee; monkey;
    orangutan
psychology, 65, 136, 141–42, 187; ADHD, 65;
    bio-, 141; embodied, 88; PTSD, 84–89,
    162; shared fundamentals of, 3, 5, 85–86,
    94, 178, 179–80; social, 141. *See also*
    attachment: style; brain; internal felt
    sense of life; mind
psychoneuroimmunology, 192
PTSD (post-traumatic stress disorder),
    84–89, 162
puma, 132

qualia, 142

rabbit, 169
raccoon, 1, 11, 46, 166
rage. *See* affective systems
Ram, Ashish, 74

rat, 119, 193

rattlesnake, 47, 67

raven, 2–8; anatomy of, 3–5; at Tower of London, 7. *See also* Jason

rehabilitation, 6, 7, 14–15, 70, 166, 180

reptile, 86, 119, 129, 130, 198

rescue. *See* animal rescue

rhinoceros, 43, 77, 83–90, 92; anatomy of, 51, 85; with BFF, 51, 58; big five, as one of, 40; horn of, 83–84, 89–90, 213n24; psychology of, 85–86. *See also* Ike; Phillipa

Rich, Adrienne, 153

Ringling Bros. and Barnum & Bailey, 176, 180. *See also* circus

rodent, 112, 116, 119, 121, 122, 176. *See also specific species*

Rogers, Ginger, 41

Roode, Adine, 59, 69, 73

Roode, Johann, 44

Roode, Lente, 44, 58, 61

Roode family, 43, 60–61

Rose, Naomi, 180

Rowlands, Mark, 182

Royal Society for the Protection of Birds, 130

Rumi, 136

Rusty (dog), 134

safari, 8–9, 40, 42–54 passim. *See also under* elephant

salamander, 167

sanctuary. *See* animal sanctuary

San Diego Humane Society, 117

San Diego Zoo, 162, 163

San Francisco Zoo, 172

Satyanarayan, Kartick, 178

Saving the Survivors, 84

saw-toothed grain beetle, 198

Schore, Allan, 106

scorpion, 50, 53, 58

SeaWorld, 181–82

Sebakwe (elephant), 69

Sebastian Pig, 37, 38, 39

Sebeka (cheetah), 44, 58

seeking system. *See* affective systems

sensory system, 80, 82, 86

sentience, 22, 56, 192, 193

Shanthi (elephant), 164

sheep, 20, 22, 32–33, 37, 60, 109; Louise, 20, 32

Sherry, Lady Bright III (dog), 151–54

Shubin, Neil, 83

Simpson, Jean, 170–71

Singer, Peter, 74

skunk, 166

Skye (goat), 20, 32, 37, 39

skylark, 130

Slobodchikoff, Con, 172, 178–79

slug, 194, 196

Smithsonian National Zoological Park, 164

snail, 193, 196–98

snake, 27, 166; black mamba, 46–47, 50–51; rattlesnake, 47, 67

social bond, 36, 55–56, 118, 162; of elephant, 62, 70–71; of human and dog, 135–36, 146, 151–54; of human and genet, 46; of human and whale, 92–110 passim

social neuroscience, 36, 191–92, 213n13 (chap. 5)

songbird, 4, 15, 119, 121, 129–30

South Africa, 9, 41, 44, 89

spider, 195, 210; Mediterranean brown recluse, 195–96

squirrel, 1, 2

Stanford University, 172

Stella (cat), 111–12

Stewart, Jon, 33

Stewart, Tracey, 33

St. Francis of Assisi, 184

Stray Cat Alliance, 129

Strutt, John, 155–56, 157–59; at Harrow, 158, 159

substance P, 139, 143, 153

Sullivan, Annie, 138

Tansey, Mark, 183

Tautz, Jürgen, 191

Taylor, Jessica, 122, 127

termite, 50, 53

Thomas, Elizabeth Marshall, 116, 132, 164

Three Rivers Humane Society, 148–50

tick, 195; Asian long-horned, 195

tiger, 180

Tigere (elephant program director), 69–70

Tilikum (whale), 181–82

Timisia (elephant), 70

TNR (trap-neuter-return), 119–32 passim

Tokwe (elephant), 61, 62–63, 69

trap-neuter-return (TNR), 119–32 passim

turkey, 22, 26, 32

UNESCO, 95
United States: agricultural landscape of, 199; Constitution, 181; Department of Agriculture, 166; insects of, 198; predation of birds in, 129; sanctuaries in, 180
University of British Columbia, 120
University of Chicago, 120, 123; Divinity School, 124, 127
University of Winchester, 23–25

vegan, 21–22, 34
vegetarian, 5, 21–22, 30, 34
veterinarian, 72, 83–84, 117, 151, 166, 213n11; Dr. Vicky, 137–40, 143; TNR, 119–32 passim
Viorst, Judith, 215n6

Walker, Alice, 28
Warm Springs Indian Reservation, 151
warthog, 9, 45, 50, 58, 61, 67; in mongoose den, 53
wasp, 195
water, 1–3, 11, 19, 53; watering hole, 55, 68. *See also* drought
Watson, John, 36
Weitzman, Gary, 117
whale, 80, 87, 130, 181, 184, 209; blue, 107; Declaration of Rights for Cetaceans, 182; International Agreement for the Regulation of Whaling, 96; orca, 103, 181–82; Tilikum, 181–82. *See also* eastern North Pacific gray whale; hunting; International Whaling Commission

Whittier, John Greenleaf, 206
wildebeest, 45
Wildlife and Countryside Act, 158
Wildlife and Environmental Society of South Africa, 44
Wildlife SOS Elephant Conservation and Care Centre, 178
Williams, Robin, 173
Wilson, Edward O., 168, 208
Winnicott, Donald, 192
wolf, 56, 87, 135–36, 171, 172; with St. Francis at Gubbio, 184
woodlouse, 194
World Heritage Site, 93, 95
World Wildlife Fund, 71
worm, 189, 198; corn, 198; earthworm, 194
Wynne, Clive, 135

zebra, 9, 45, 67
Zimbabwe, 60, 61, 67
zoo, 160, 162–64, 170, 176, 180; animal enrichment in, 162, 163–64, 166; animal release from, 180; big cats in, 164–65; elephants in, 62, 161; ethics of, 162–63, 164–65; mental illness in, 161, 162; Minnesota Zoo, 161–62; National Aquarium, 180, 182; San Diego Zoo, 162, 163; San Francisco Zoo, 172; Smithsonian National Zoological Park, 164. *See also* CALM

ANIMAL VOICES, ANIMAL WORLDS

ERIN MCKENNA, *Livestock:*
Food, Fiber, and Friends

ARNOLD ARLUKE AND ANDREW ROWAN, *Underdogs:*
Pets, People, and Poverty

MICHELE MERRITT, *Minding Dogs:*
Humans, Canine Companions,
and a New Philosophy of Cognitive Science

ANNE BENVENUTI, *Kindred Spirits:*
One Animal Family